DENYING THE WIDOW-MAKER

Summary of Proceedings
RAND-DBBL Conference on Military Operations on Urbanized Terrain

Russell W. Glenn
Randall Steeb
John Matsumura
Sean Edwards
Robert Everson
Scott Gerwehr
John Gordon

With
Fenner Milton
Timothy Thomas
Randall Sullivan
Tony Cucolo
Gary Schenkel

Prepared for the United States Army

ARROYO CENTER

RAND

T0146077

For more information on the RAND Arroyo Center, contact the Director of Operations, (310) 393-0411, extension 6500, or visit the Arroyo Center's Web site at http://www.rand.org/organization/ard/

On February 24–25, 1998, the RAND Arroyo Center and U.S. Army Infantry Center's Dismounted Battlespace Battle Lab (DBBL) co-hosted a conference on Military Operations on Urbanized Terrain (MOUT) at the RAND Washington, D.C. office.

The conference had three primary objectives:

- Assess the current state of U.S. MOUT readiness and determine viable approaches to any changes deemed necessary during the periods 1998–2005 and 2006–2025.

- Provide a forum for debate on MOUT issues, including

  — discussion of new operational concepts,

  — analysis of training techniques and procedures,

  — determination of high-resolution simulation requirements.

- Determine what technologies are worthy of consideration and high-acquisition priority under the auspices of Advanced Concepts Technology Demonstration (ACTD), Force XXI, and Army After Next (AAN).

The initial foundation for discussions was formed by two questions:

- Should there be a dramatic change in the U.S. military's concept for MOUT?

- If so, when is such a change practical and how should it take place?

A list of persons attending the conference will be found in Appendix B.

The persons listed "with" the authors of this report provided the slides corresponding to their conference briefings (Those slides appear in Appendixes C–H). The listing of their names does not constitute their endorsement of this document.

Research conducted in conjunction with this document was carried out in the RAND Arroyo Center, a federally funded research and development center sponsored by the United States Army. This work was conducted in the Force Development and Technology Program for the Assistant Secretary of the Army (Research, Development and Acquisition) and the U.S. Army Deputy Chief of Staff for Operations.

This summary of proceedings will be of interest to government and commercial-sector personnel whose responsibilities include policy design, funding, planning, preparation, or the development of technologies for, or the conduct of operations in, urban environments.

CONTENTS

Part 1: DESCRIPTION OF PROCEEDINGS

Part 2: APPENDIXES

TABLES

Recent months have increased our awareness of the potential challenges posed for the United States by urban areas worldwide. Several factors—the coincidence of U.S. military superiority; commitment of the nation's ground forces to contingencies in cities such as Panama City, Mogadishu, and Port-au-Prince and the continued presence of its men and women in Bosnian built-up areas; sustained, perhaps even increased, concerns over American and indigenous noncombatant casualties; and the inherent difficulty of operating in urban environments—have sparked the recognition that limited preparedness threatens the country's national interests.

This document is a summary of a two-day conference on military operations on urbanized terrain (MOUT) held at the RAND Washington, D.C. office on February 24–25, 1998. The conference was co-hosted by the RAND Arroyo Center and the United States Army Infantry School Dismounted Battlespace Battle Lab (DBBL). The agenda included presentations on recent historical events (Grozny, Hue), ongoing operations in urban areas (Bosnia), and initiatives under way to improve future force readiness to conduct military operations in cities. Afternoon sessions challenged conference participants to develop near- and longer-term approaches to attaining such improvements. This summary gathers the views presented and the issues debated during the two days. Copies of the slides used by conference speakers appear in the appendixes.[1]

---

[1]LTG George R. Christmas spoke of lessons to be learned from fighting in 1968 Hue, Vietnam, and of the human factors issues inherent in urban combat. His presentation did not include informational slides, thus none appear in the appendixes.

Two assumptions underlay the participants' discussions of issues during the two-day session:

- The United States will no longer conduct urban operations without significant rules of engagement (ROE) designed to reduce noncombatant casualties and collateral damage to infrastructure.

- National decisionmaking authorities will continue to perceive the American public as intolerant of large numbers of U.S. military casualties in any but exceptional circumstances.

Conference attendees agreed that a continued reliance on World War II–type combat methods for operations in cities was counterproductive. While it was recognized that near-term improvements would be limited to enhancing current procedures via modified doctrine, training, and extant or proven concept technologies, such changes could at best result in marginal upgrades in force readiness. For the longer term, alternatives to large-scale commitments of U.S. manpower into urban areas and subsequent engagement of adversaries at close range was deemed desirable. Early research indicates that such a significant change in methodology may be feasible by the opening years of the next century's third decade.

Conference attendees reviewed the current status of U.S. MOUT readiness and analyzed potential improvements within a framework of four categories: doctrine, training, organization, and technology. Changes in the first two, doctrine and training, were believed to hold the greatest potential for near-term improvements; but significant modifications to both are essential to attain such improvements. Most attending the conference concluded that U.S. military organizations should remain capable of conducting a wide variety of missions and thus that creation of any environment-specific units (e.g., a MOUT-specific division) was ill-advised. Even those supporting possible changes to organizations believed that execution of changes should be delayed until test structures could be evaluated in Marine Corps and Army exercises.

Changes to doctrine and training can have a significant effect in improving force capabilities to conduct MOUT and in reducing friendly and noncombatant casualties while so doing. However, there are definite limits to such enhancements. Only with the emergence of new technologies to accompany innovative doctrinal and training approaches can dramatic increases in capabilities be achieved. Though the specifics of a new approach to MOUT were left undetermined by those attending the conference, concepts and technologies worthy of further consideration were identified.

ACKNOWLEDGMENTS

The authors express their appreciation to the speakers and attendees at the February 1998 RAND-DBBL MOUT conference. The stimulus provided by the speakers' superb presentations established the foundation for two days of vigorous free exchange on the subject of military operations in urban areas. Those in attendance are to be commended for walking the fine line that allows no-holds-barred challenges of others' views while never failing to maintain an atmosphere of professional collegiality.

Special thanks are due to RAND's co-hosts of the event, the men and women of the United States Army Infantry School Dismounted Battlespace Battle Lab and those affiliated with that organization. Notable in their support were Colonel Timothy G. Bosse, Lieutenant Colonel Randall Sullivan, and Carol Fitzgerald.

The success of the conference was in very considerable part attributable to a woman whose behind-the-scenes efforts ensured that the affair went flawlessly. Our thanks to Donna Betancourt, Director of Operations, Army Research Division, RAND Arroyo Center.

The timely fashion in which this document was released is entirely attributable to two highly skilled and extraordinarily dedicated women, Patrice Lester and Terri Perkins.

We also recognize the professionalism and dedication of this document's reviewers. Robert Howe of RAND and Lieutenant General Ron Christmas (USMC, ret.) both provided valuable observations and comments of service to our readers.

We thank the following individuals and organizations for their permission to use photographs as displayed in Appendix F: AeroVironment Inc., Simi Valley, CA (micro-UAV); Space and Naval Warfare Center, San Diego, CA (MDARS-E); Northrop-Grumman (Andros); and Peter Bialas, KETIC Inc. (Seoul).

ACRONYMS

AAN	Army After Next
ACTD	Advanced Concepts Technology Demonstration
C4ISR	Command, Control, Communications, Computers, Intelligence, Surveillance, and Reconnaissance
CINC	Commander in Chief
DBBL	Dismounted Battlespace Battle Lab
FEBA	Forward Edge of the Battle Area
FM	Field Manual
HUMINT	Human Intelligence
IFF	Identification, Friend or Foe
JRTC	Joint Readiness Training Center
JTF	Joint Task Force
LTC	Lieutenant Colonel
Lt Col	Lieutenant Colonel (British)
LTG	Lieutenant General
MCWP	Marine Corps Warfighting Publication
MOBA	Military Operations in Built-up Areas
MOUT	Military Operations on Urbanized Terrain

NCA	National Command Authorities
OCODA	Observe, Culturally Orient, Decide, Act
OODA	Observe, Orient, Decide, Act
OPFOR	Opposing Forces
ROE	Rules of Engagement
RPG	Rocket Propelled Grenade
SFC	Sergeant First Class
SOF	Special Operations Forces
USMC	United States Marine Corps

Part 1
Description of Proceedings

Chapter One

INTRODUCTION

SUMMARY OF ISSUES

The scope of experience and expertise represented by those who attended the RAND-DBBL MOUT conference ensured a similar breadth in approaches to and concerns about the subject area. This section first provides an overview of issues raised and then addresses them more specifically within a framework of four topic areas: doctrine, training, organization, and technology.

OVERVIEW

Conference attendees were in general agreement on the shortcomings of current U.S. armed forces' MOUT capabilities and the potential disconnect between these capabilities and successful service of national interests. Several areas were of particular concern at the macro level:

- Doctrinal guidance regarding avoidance of urban operations,

- Effects of casualties on national decisionmakers and military mission accomplishment,

- The unfeasibility of immediate solutions to MOUT shortfalls and the likelihood that urban operations will be essential in the near term.

Individuals and groups proposed several recommendations for addressing these concerns. One notable bright spot was the recognition that many steps taken to improve force readiness to conduct urban operations would have positive influences on preparedness for activities in other environments, a significant point in times of constrained economic and manpower resources.

There were widespread concerns that current doctrine recommends avoidance of urban operations when demographic trends make avoidance an unlikely alternative. This contradiction between guidance and potential commitments was thought to have a number of negative implications. First, acceptance of an avoidance policy, it was believed, would reduce the likelihood of MOUT receiving high priority for funding and improvements in doctrine and training: if avoidance is the accepted norm, a logical setting of priorities would place emphasis on readiness involving scenarios where operations were more acceptable and thus more likely. Second, this doctrinal guidance was inconsistent with recent National Command Authorities' directives, e.g., committing military forces to urban operations in Panama, Somalia, Haiti, Liberia, and Bosnia. Avoidance, while remaining a *desired* alternative, had to be recognized as an increasingly unacceptable or unavailable one.

Third, anxiety in this regard was complemented by recognition that MOUT success might be beyond the threshold of the possible in selected contingencies and that current doctrine fails to recognize these potential limits. The size of many modern cities is such that traditional missions of seizing and controlling them are probably beyond the ability of the U.S. military today and in coming years, even when our forces are engaged as members in an alliance or coalition. The largest cities in which U.S. forces have conducted large-scale, high-intensity combat operations (e.g., Manila in 1945, Seoul in 1950) had populations in the neighborhood of one million; their populations and geographic expanse were far less than those of major metropolitan areas today. Determination of MOUT capability thresholds and development of alternatives to seizure of entire cities were considered essential steps toward developing an understanding of feasible U.S. military courses of action during future contingencies.

The historical relationship between high casualties and urban operations was a second area of concern. The vulnerability of U.S. operations to friendly casualties was demonstrated by the decision to remove the nation's forces from Somalia after the October 3–4, 1993, fighting that resulted in 18 Americans killed and another 73 wounded. There is evidence that noncombatant casualties can also influence political and military decisionmaking; the modification of bombing policy after the Al Firdos bunker incident in the 1991 Persian Gulf War is a recent example. The extent of noncombatant casualties' influence on U.S. policy is unclear, however, and stands as a topic requiring further analysis. Nevertheless, the historically high cost of urban combat operations, and the

> *When we take into account the perceived benefits of the operation. . . the evidence of a recent decline in the willingness of the public to tolerate [friendly force] casualties appears rather thin. The historical record in fact suggests a rather high degree of differentiation in the public's willingness to tolerate casualties based upon the merits of each case. Whenever the reasons for introducing U.S. forces lack either moral force or broadly recognized national interests, support may be very thin indeed, and even small numbers of casualties may often be sufficient to erode public support for the intervention.[2]*
>
> —Eric Larson

apparent intolerance for friendly and noncombatant loss of human life, poses an additional problem for a military force confronted by an urban contingency that is critical to national interests. Table 1 provides some sense of these costs.[1]

Finally, the potential for U.S. armed forces commitments to combat MOUT in the near term and the inability to immediately upgrade readiness for these operations troubled conference attendees. Most concluded that though the nation's MOUT readiness can be improved in the near term, resources and technological limitations will preclude its being "fixed" during this period. In 20 to 25 years, however, U.S. forces could develop ways to conduct urban combat operations that would significantly reduce both friendly and noncombatant casualties. Those attending the conference could not agree

---

[1]During his conference briefing, Russell W. Glenn provided a nontraditional definition of operational success. Military mission accomplishment, it was posited, is of itself no longer sufficient. Rather:

Success = Military mission accomplishment + Reasonable friendly force casualties + Tolerable noncombatant casualties

[2]Eric Larson, *Casualties and Consensus: The Historical Role of Casualties in Domestic Support of U.S. Military Operations,* Santa Monica, CA: RAND, MR-726-RC, 1996, pp. 99–100.

on the exact nature of this longer-term fix. A RAND Arroyo Center proposal involved removing U.S. soldiers and marines from close combat in any but exceptional circumstances and providing them with the means to execute precision engagements from remote locations. Other representatives suggested that some (undetermined) U.S. "overmatch" capability was key to attaining success.

Several additional general recommendations were developed to improve both short- and longer-term readiness for urban operations. There was widespread agreement that service chiefs, CINCs, members of Congress, and the National Command Authorities (NCA) in general needed to upgrade the attention given to MOUT issues. Attendees were concerned that the absence of an affiliation of MOUT requirements with a "big ticket" system made urban initiatives unattractive for funding at the requisite level. Service chief and CINC emphasis was deemed essential for the development and maintenance of viable technological, doctrinal, training, and operational initiatives; without champions, MOUT will remain an area of indi-

Table 1

**Friendly and Noncombatant Deaths in Recent
U.S. Urban Combat Operations**

Battle	Noncombatant Killed in Action[a]	U.S. Killed in Action	Noncombatant : U.S. Killed in Action ratio (approximate)
Manila (1945)	100,000 (estimated)	1,010	100:1
Hue (1968)	5,800	150[b]	40:1
Panama (1989)	202	26	8:1
Mogadishu (1993)	500+	18	30:1

[a]Noncombatant deaths in these battles were not all the result of U.S. fires. In Hue, for example, South Vietnamese citizens executed by Communist forces are estimated at some 2,000. However, the table does reflect the high cost of urban warfare for a city's residents.

[b]In addition to the U.S. casualties, 400 soldiers from the Army of the Republic of Vietnam lost their lives.

vidual service and joint neglect until an operational shock of sufficient magnitude forces it to the forefront.[3]

Further recommendations pertained to service conduct of urban operations. There was a call for a more disciplined approach to future MOUT as opposed to what some characterized as the current "brilliant innovation" methodology, i.e., waiting for problems to arise and addressing them via on-the-spot solutions. It was also concluded that the services should include strategic, operational, and tactical implications in their preparations for and conduct of MOUT rather than focusing exclusively on the tactical level.[4] Technology, while a critical component of such analyses, was thought by many to be overemphasized to the detriment of doctrinal, training, organizational, and other issues. Specific areas thought to have been heretofore neglected or underemphasized included psychological operations, deception, and stress effects on friendly force members. Finally, there was recognition that MOUT encompassed the complete spectrum of military operations, not exclusively combat actions; any consideration of the subject must recognize the scope and heterogeneity of its demands.

---

[3]The USMC arguably has had two recent MOUT champions: the Commandant of the United States Marine Corps and LTG Van Riper at Quantico. MG Ernst's work on urban operations at the U.S. Army Infantry School and Center is both critical and significant, but the Infantry Center MOUT focus is primarily brigade and below; any Army-wide or joint champion has to include all levels of operations in his scope of interest, have a combined arms and joint (and preferably multinational and interagency) perspective, and wield sufficient influence to ensure the subject is not ignored. The 1994 Defense Science Board likewise called for a MOUT champion in its report. See Office of the Under Secretary of Defense for Acquisition and Technology, *Report of the Defense Science Board Task Force on Military Operations in Built-up Areas (MOBA)*, November 1994, p. 2.

[4]However, a minority believed that an urban area is essentially a small-unit-action environment and that any efforts to enhance U.S. capabilities should focus on that level, particularly in the area of small-unit leader development.

Chapter Two

DOCTRINE

> The five-month battle in and around Stalingrad . . . was only a part of a far vaster drama played across an immense stage of steppes and forests and mountains. . . . As stiffening Soviet resistance, plus their own difficulties, slowed the German advance, Hitler milked more and more troops away from the vital northern flank . . . the hinge of the whole operation . . . to re-enforce Paulus's 6th Army at Stalingrad.[1]
>
> —*The Battle for Stalingrad*

The role of doctrine in establishing a foundation for MOUT training, technological development, and potential organizational change was widely recognized, as was the fundamental necessity for a doctrine that includes consideration of more than the tactical level of operations. There was a call for revision of current doctrine at all levels with an emphasis on maintaining continuity from strategic-level guidance to that provided for the individual soldier and marine. This call also stressed the need for joint MOUT doctrine. Both joint and service doctrine need to be comprehensive not only in the sense of the levels of operations, but also with regard to the complete spectrum of potential operations (e.g., stability and support missions as well as those entailing combat), multinational issues, and interagency considerations. There was some unresolved debate between a minority who believed the primary focus of MOUT doctrine should be at the tactical level and those who posited that current doctrine already had too great a tactical focus.

---

[1]Hanson W. Baldwin, in the introduction to Vasili I. Chuikov, *The Battle for Stalingrad*, New York: Holt, Rinehart, and Winston, 1964, p. 3.

Several strategic and operational elements were thought to be essential to the creation of effective doctrine. The responsibilities of CINCs for making requirements known, overseeing training, promoting technological development, and supporting other activities critical to MOUT preparedness should be identified. A definition of joint "MOUT" and analysis of current and future cooperative approaches to urban operations was considered necessary. The difficulties confronting air and aviation elements during MOUT raised questions on whether joint air operations in MOUT required specialized tactics, techniques, procedures, and command relationships. Several individuals felt there was inadequate description of how joint operations should be conducted during urban contingencies. Effective operational-level doctrine was thought to be fundamental for guiding Joint Task Force (JTF) operations and training. None exists at present; current joint MOUT doctrine was seen as little more than "lip service."

Another important element of this doctrinal enhancement will be the addressing of the limits of U.S. capabilities. Joint doctrine must account for diminished force strengths; no longer can the United States expect to commit multiple divisions to a long-term urban operation unless means of dramatically reducing personnel losses are developed. A precipitate conclusion is that doctrine to achieve desired end states via methods other than complete seizure or clearing of a built-up area is necessary.

Doctrine should emphasize that urban operations may not be desirable for many reasons, but that a policy of avoidance is no longer viable in many circumstances. Given the alternative of fighting in a city or doing so elsewhere, guidance to select the second is probably wise. Increasingly, however, no such alternative will be available; it is therefore essential to synchronize strategic expectations and capabilities. To paraphrase Clausewitz, any representative of the National Command Authorities must be "aware of the entire political situation," while also understanding "exactly how much he can achieve with the means at his disposal."[2]

---

[2]Carl von Clausewitz, *On War* (Michael Howard and Peter Paret, eds. and trans.), Princeton, NJ: Princeton University Press, 1976, p. 112.

A new doctrine must include the complete scope of likely military activities. The USMC has recently labeled the possibility of humanitarian missions, stability activities, and combat occurring simultaneously during a single operation and in a single city as the "three block war." This contingency, in which units have three dramatically different types of actions ongoing over distances measured in a few hundred meters, is virtually unrecognized in doctrine today.

It was recommended that future MOUT doctrine take a broader and more integrated approach to address these varied demands. Guidance for combat operations should not be neglected, but doctrine to cover other scenarios was deemed essential. For example, manuals should include discussions of how to select and neutralize critical urban nodes to facilitate success during stability missions. Similarly, coverage of how to prioritize and restore essential services is necessary. Doctrine should cover contingencies such as those now commonplace in Bosnia, situations in which soldiers and marines must constrain their actions to meet stringent rules of engagement but be prepared for the high-intensity MOUT that could be but "one heartbeat away."[4] Dealing with such divergent requirements requires guidance that considers the use of both lethal and nonlethal means of engagement.

> *Rules of engagement required precision fire, used only when absolutely necessary, particularly from the powerful Spectre gunship and the Rangers once they were heading toward their objectives.*
>
> *The principle handed down by the Just Cause commanders to minimize enemy casualties and collateral damage was relentless. "We would carpet bomb the airfield if we could" [Colonel William Kernan, Commander, 75th Ranger Regiment said]. But the Spectre was only allowed to fire on key military targets that threatened or had the potential to affect the Rangers' mission.*[3]
>
> *—Thomas Donnelly, et al.*

---

[3]Thomas Donnelley, Margaret Roth, and Caleb Baker, *Operation Just Cause: The Storming of Panama*, New York: Lexington, 1991, p. 193.

[4]The quotation is taken from a conference participant who used it to describe the thin line that sometimes exists between combat and a noncombat environment during stability and support missions.

There is an immediate need to update joint publications and the Army's Field Manual (FM) 90-10, *Military Operations on Urbanized Terrain*.[5] Authors of subordinate manuals rely on those documents as the basis for development of supporting doctrine.

Those attending the conference generally agreed that although improvements in the current approach to MOUT (designated as "close combat MOUT" due to its reliance on closing with an adversary and engaging him at short range) might mitigate friendly casualties to a limited extent, success in addressing today's strategic requirements may not be attainable unless an alternative method is adopted. However, there was no universal agreement on the nature of that alternative. The aforementioned RAND strategy designed to remove U.S. soldiers and marines from an environment in which short-range engagements were the norm was believed by some to be both desirable and feasible for the long term. Given a commitment to move toward this capability, steps to capitalize on advances in that direction should be taken when possible. Others were unconvinced that this was the proper approach, but no other predominant concept emerged. It was agreed, however, that future doctrine should support reductions in force vulnerability and the employment of alternatives to the commitment of friendly forces in high-risk situations when other viable courses of action are available.

The following additional areas were noted by one or more persons as requiring further attention in MOUT doctrine:

- **Fire support:** Both fire support doctrine and the training derived from it were seen as deficient in the urban operations arena.

- **Casualty evacuation:** Current reliance on limited numbers of medical personnel results in combat personnel having to assist with medical evacuation, further draining fighting strengths in an environment notorious for manpower consumption.

- **Intelligence collection, analysis, and dissemination:** The short duration of response times and the fleeting nature of opportunities demand rapid and accurate completion of the

---

[5]The Marine Corps Warfighting Publication (MCWP) 3-35.3, *MOUT,* was being reviewed and updated at the time of the conference. The document was released in April 1998.

collection → analysis → dissemination intelligence cycle. MOUT are squad leaders' operations to a large degree, so promulgation of real-time intelligence must be to the lowest levels if they are to be of value.

- **Psychological operations:** Chechens' effective use of psychological operations against their Russian adversaries emphasized both the value of well-conceived psyops and the need to prepare friendly forces for an adversary's employment of this resource.

- **Early deployment of human intelligence (HUMINT), psychological operations, information operations, and deceptions assets into theaters involving MOUT:** Preliminary identification of critical nodes, preparation of the indigenous population, determination of optimum insertion points, establishment of contacts with local guides, and other activities crucial to success and requiring significant lead times are difficult or impossible to accomplish effectively if initiation is too greatly delayed.

- **Defensive MOUT:** Doctrine for defensive urban combat operations is very limited. Future doctrinal improvements should adequately address friendly and enemy defensive operations to include the use of "layers" or "belts."

- **Anti-tank defense:** Techniques to better deny an enemy effective employment of RPGs and use of defensive positions above or below the depression limits of IFV and tank guns were seen as essential.[6]

- **Deception:** The conference attendees called for doctrine that (1) supports seizing opportunities offered by deception, and (2) ensures superior friendly C4ISR operations.

- **Command and control operations:** Russians in Chechnya modified the well-known OODA (Observe, Orient, Decide, Act) loop to account for cultural differences confronted during operations in the city of Grozny, labeling it an OCODA (Observe, Culturally Orient, Decide, Act) loop instead. U.S. forces must be

---

[6]The significance of RPGs in MOUT is evident from their status as the most feared weapon during Russian fighting in Grozny (Timothy L. Thomas, lecture at RAND-DBBL MOUT conference, February 24, 1998) and the vulnerability of circling helicopters to their fires in 1993 Mogadishu.

ready to operate in built-up areas where communications are degraded and the pace of decisionmaking is increased due to the close proximity and density of forces.

> *I remembered battles against the White Guards and White Poles in the [Russian] Civil War when we had to attack under artillery and machine-gun fire without any artillery support of our own. We used to run up close to the enemy, and his artillery would be unable to take fresh aim and fire on rapidly approaching targets. . . . I came to the conclusion that the best method of fighting the Germans would be close battle, applied night and day in different forms. We should get as close to the enemy as possible so that his air force could not bomb our forward units or trenches.[7]*
>
> —*Vasili I. Chuikov*

- **Enemy hugging tactics (as used by the Russians in Stalingrad):** These were envisioned as a likely response of an adversary to U.S. targeting capabilities. There is therefore a need to address countermeasures.

- **Clearing techniques, to include how to select and prioritize structures to be cleared:** There is a requirement to develop doctrinal procedures for clearing rooms and structures that (1) best meet mission requirements, (2) minimize danger to friendly force members, and (3) preclude unnecessary noncombatant casualties.

---

[7]Vasili I. Chuikov, *The Battle for Stalingrad,* New York: Holt, Rinehart, and Winston, 1964, p. 72.

Chapter Three

TRAINING

Grozny was cited as an example in which a technologically more sophisticated force fell prey to a less well equipped but better disciplined and trained one. The lesson to be learned for the United States is that MOUT requires the dedication of more training time than is currently the norm in the nation's armed forces.

Obstacles to improved training are significant. Most units assign a low priority to preparing for operations in built-up areas. A British representative bemoaned the fact that units in his army spent only 8 percent of their training hours on the subject. The time spent by most American units is far less. Preparation is further hindered by a lack of training facilities. Those MOUT training sites that do exist are few and small in size. Light force units are to complete a cycle through the Joint Readiness Training Center (JRTC) at Fort Polk once every 18–24 months, a rotation that includes training at that installation's MOUT complexes, but these organizations often have little other training in built-up environments. Only five of the eight active Marine Corps regiments have a local MOUT site on which to train.

In addition to this lack of dedicated training time and facilities, nearly all MOUT training is completed only at the small-unit (battalion and below) level. U.S. urban training facilities are too small for brigade or larger operations; most are of such size that even platoons or companies exceed a facility's capability to provide an attacking unit with a viable threat to its flanks from within the built-up area. The need for far more extensive training sites was a topic of considerable discussion during the conference. Several in attendance recommended the creation of a National MOUT Training

Center, one preferably designed and run as a joint training site. Such a facility would ideally be capable of handling training for a minimum of a brigade-sized unit comprising both heavy and light forces. It would be fully instrumented; capable of integration into an operational-level scenario; include both force-on-force and live-fire training; and integrate noncombatant, multinational, and interagency considerations. Such a large facility should include simulations to assist units in enhancing MOUT skills during their rotations at this joint facility. The possibility of using a post designated for closure or an unused shipyard as such a training site was forwarded as a possibility.

Whether they are co-located in a national MOUT center or elsewhere, there is a need not for a single training facility but for several that together account for the heterogeneity in types of built-up environments. It was suggested that training for urban operations

> Fighting inside Hue involved a kind of experience that neither [officer] possessed. Nor had any of their young troops ever been trained in city fighting.[1]
>
> —Eric Hammel

should include exposure to small villages (akin to current MOUT facilities), medium-sized towns or small cities, and megalopolises. In the last case, use of actual cities would likely be the only feasible alternative.

Most participants concluded that MOUT training should have combat operations as the primary focus. Units designated for operations consisting primarily of stability and support missions would receive supplemental training before deployment, much as is done in the British Army before units are sent to Northern Ireland.[2] One critical element of this preparation would be to train soldiers and marines in the application of what one conference speaker called "the switch," an instantaneous transition from one type of behavior and mind-set

---

[1]Eric Hammel, *Fire in the Streets: The Battle for Hue, Tet 1968*, New York: Dell, 1991, p. 97.

[2]British Army units generally train for three months before a six-month Northern Ireland tour.

to another as demanded by the situation.[3] Use of aggressive but nonlethal force to demonstrate resolve, displays of consideration for civilians in the operational area, and a readiness to switch to the high-intensity MOUT that may be but "one heartbeat away" are all part of operations involving stability and support missions. Such discipline comes only through extensive and well-designed training.

Other identified training shortcomings and related recommendations included the following:

- Introduce MOUT exercises to Command and General Staff College and War College training.

- Provide guidance for baseline training certification.

- Include changes in rules of engagement (ROE) during training and exercises. ROE changed significantly during fighting in both 1968 Hue and 1995 Grozny.

- Use actual towns and cities for routine MOUT training, e.g., leader terrain walks.

- Provide dedicated OPFOR and noncombatant personnel at urban training facilities.

---

[3]Tony Cucolo used the term during his presentation "MOUT in Bosnia: Experiences of a Heavy Task Force, December '95 to December '96." See Appendix G.

Chapter Four

ORGANIZATION

In our counter-attacks we abandoned attacks by entire units and even sections of units. Towards the end of September storm groups appeared in all regiments; these were small but strong groups, as wily as a snake and irrepressible in action.[1]

—Vasili I. Chuikov

The prevailing recommendation was to retain a flexible, generic organizational structure rather than create MOUT-specific organizations. It was felt that current task-organizing procedures provided sufficient flexibility for individual service and JTF contingencies. Creating specially designed units was thought to be counterproductive, as all forces should be capable of operating in any conceivable environment. However, some thought it might be feasible and perhaps advisable to raise the MOUT proficiency of selected units, perhaps one battalion per brigade, to create a force capable of quickly training compatriot units. Even the small minority that considered the creation of MOUT-specific organizations as a legitimate option recommended not making any changes until after testing possible alternatives during Urban Warrior and Army After Next exercises.

Though redesign of units for MOUT purposes was rejected, several changes related to force structure were considered necessary. An increase in human intelligence collection, psychological operations, civil affairs, and information warfare effectiveness was thought advisable due to the inevitability of having to deal with large numbers

---

[1]Chuikov, *The Battle for Stalingrad*, p. 146.

of noncombatants and other demands of urban missions.[2] Maintenance of a robust SOF capability was likewise deemed essential due to the special skills those organizations bring to fighting in cities. Some believed that selected urban operations tasks could be assumed by contract organizations rather than military units. Reduced manning in infantry units was thought to present a degradation of critical active component operational capabilities.

---

[2]Timothy Thomas noted the significant role played by psychological operations and HUMINT during Russian operations in Grozny.

Chapter Five

TECHNOLOGY

While there was widespread recognition that any effective future enhancement of U.S. MOUT capabilities will require a combination of advances in doctrine, training, technology, and perhaps organization, many at the conference warned against an overreliance on technology as a solution. Technology was not envisioned as a significant factor in the near-term enhancement of U.S. MOUT readiness. It was recognized that although selected technologies proffered benefits, their impact would be less than those obtained via changes to doctrine and training. This was in considerable part attributable to two factors: (1) the belief that urban operations tend to mitigate technological superiority, as was the case in 1993 Mogadishu and 1995 Grozny, and (2) a conviction that technologies likely to be available to the field in the immediate future would not dramatically alter the character of nor losses resulting from U.S. MOUT operations. For the near term, then, many attendees felt time, effort, and funds would be better spent in improving doctrine and training; technology should not be "oversold" as a solution to the challenges of MOUT.

Despite these caveats, the potential near-term benefits that technology does offer should not be overlooked. The joint USA/USMC ACTD and USMC Urban Warrior initiatives recognize that development of specific technologies offer significant combat multiplier capabilities. Application of Marine Corps urban thrust, urban swarm, and urban penetration techniques will demand a quality of intelligence and coordination, speed of movement, and precision engagement impossible without production of systems essential to their support.

There were, however, specific concerns about U.S. technological development. Several thought that weapon systems and other innovations were being adopted without sufficient attention being paid to their applicability during urban operations. The same individuals expressing such concerns, however, were quick to warn that similar systems should no more be specially designed for MOUT than should others be brought into the force structure without proper consideration for applications in built-up areas. There was also a call for determining the legal and political implications of adopting acoustic, microwave, laser, and nonlethal chemical capabilities.

Speaker presentations both confirmed historical difficulties presented by urban operations and offered insights into what systems were notably valuable during recent contingencies. Russians fighting Grozny's Chechens had the same communications problems and difficulties with GPS that have plagued U.S. and other nations' armed forces. A weapon found to be especially helpful was the Russian flame-thrower, a system capable of engaging targets at ranges of up to 1,000 meters. Rather than the traditional system that fired a constant stream of flame, the weapon used against the Chechens had munitions that burst into flame on impact, with effects similar to those of fuel-air explosives. The collateral effects were not discussed.

Other technology requirements noted either by speakers or during afternoon exercise sessions:

- A system linking intelligence collection, analysis, distribution, and engagement nodes that would allow timely precision engagement. Components of this system should include (but not be limited to) digital mapping; an Identification, Friend or Foe (IFF) capability; decision-assisting software and simulations; and an enemy-noncombatant discrimination mechanism. An analysis and visualization system capable of modeling cascading disruptions or "knock-on" effects of neutralizing selected nodes would be a valuable planning tool in this regard.

- Variable effects munitions, e.g., "dial-a-yield" rounds allowing varied effects or trajectory characteristics. Such munitions would include nonlethal capabilities.

- Survivability enhancements, to include

— Near-term availability of systems to improve soldier survivability that include protection from chemical or biological
threats while not hindering mobility.

— Rapidly emplaced defensive operations materials for use
during MOUT.

— A means of protecting vehicles from weapon systems positioned beyond barrel elevation or suppression limits and
other likely threats (e.g., RPG).

— Selection (or development) of appropriate vehicles for
MOUT missions, e.g., systems other than the vulnerable systems used in Mogadishu or those such as the M113 (which
exposes the machine gun operator to fire when the weapon
is fired).

— Obscurants capable of rapid application over a wide area
with longer on-station time even in the swirling winds found
in cities.

— A stand-off bunker or wall breaching capability. Robotic
breaching, to include detection and neutralization of booby
traps, was provided as an example technology.

— A means of scanning rooms and buildings to determine if
they are occupied and a method for marking them after
scanning or clearing. Thru-wall, ground penetration sensors,
or other systems that would preclude soldiers or marines
having to enter areas for clearing, are badly needed.

• Remotely emplaced deception and other information warfare
systems, e.g., emitters providing specific signatures; jammers.

• Weapons with characteristics suitable to environments with
strict ROE and restricted maneuver space, for example, a nonlethal means of denying access to buildings or use of streets and
alleys, or low-speed, highly maneuverable fiber-optic missiles.

• Stealth-like insertion of forces, logistical support, sensors, and
other elements of combat power through the use of robots,
guided parafoils, GPS-guided munitions, or alternative means.

• Logistics enhancements:

— Effective means of resupply as far forward as the FEBA.

- — Casevac and forward care that rapidly moves casualties to the rear with a minimum burden on forward units and a maximum quality of care for wounded.

- — Reduction of the "logistics footprint" so as to minimize enemy acquisition and engagement of friendly assets.

- Simulations providing the capability to test and evaluate new doctrine and technologies. Some attending the conference noted the desirability of creating simulations designed for multiple roles, e.g., analysis, planning, training, rehearsals, and real-time visualization. However, no existent simulation is considered sufficiently mature to provide even fundamental analytic insights beyond more than a few basic issues, e.g., required sensor quality; mobility and survivability needs.

Chapter Six

CONCLUSION

> The Russian leadership did not do a good job of preparing the "theater" for warfare. The High Command did not properly seal off the republic's borders, nor did it take the time required to rehearse properly for the potential scenarios that Dudayev [the Chechen commander] had prepared for them. . . . Dudayev had created numerous firing points, communications nets, and underground command points which made the job much more difficult. In this respect, the main military intelligence (GRU) and federal counterintelligence service (FSK) did poor jobs of providing information on the illegal armed formations that the Russian force faced. . . . Russian generals not only failed to properly train their forces for combat in built-up areas . . . it became clear that the political leadership did not know how and when to use military force. [1]

> —Tim Thomas

> We should not be too quick to criticize the Russians in Grozny. . . . All of these lessons discussed by Tim Thomas were experienced in Hue.[2]

> —LTG George R. Christmas (USMC, ret.)
> Recipient of the Navy Cross for actions during
> Tet 1968 fighting in Hue, Republic of Vietnam

As has been noted, there was general though not universal agreement that continued reliance on close combat threatens the U.S. mil-

---

[1]Timothy L. Thomas, "The Caucasus Conflict and Russian Security: the Russian Armed Forces Confront Chechnya III. The Battle for Grozny, 1–26 January 1995," *Journal of Slavic Military Studies*, Vol. 10 (March 1997), p. 54.

[2]LTG Christmas made this remark during his presentation "MOUT in Vietnam-Hue, 1968."

itary's ability to meet the demands of national interests. All attending the RAND-DBBL MOUT conference recognized that the near term offers no alternative to improving the way in which the armed forces currently conduct urban operations, a manner similar to that employed in World War II, Korea, Vietnam, and since, and that such improvements are undeniably essential. Though technological advances hold some promise to assist in improving MOUT readiness in the immediate future, well-advised changes to doctrine and better training were seen as pre-eminent means of obtaining gains during this period. Looking farther into the future, however, most felt there were alternatives to maintaining a doctrinal status quo. Many considered feasible a transition to a tactical approach that would remove soldiers and marines from a norm of short-range engagements. Others were less sanguine. Yet though all recognized the need to maintain a close combat capability even in the third decade of the next century, only a few believed such methods should characterize the dominant approach. The roles of doctrine and training were seen as no less important in these out years, but technology too was perceived as a potentially dramatic complement in enhancing MOUT preparedness. Funding for MOUT technologies may suffer due to the inherent complexity of urban military operations, however. MOUT successes will be due to the synergistic effects of many factors and systems, none of which are

> Another twist unique to city fighting had become apparent in the waning hours of February 3 [1968]. Quite naturally, the 3.5-inch rocketmen had aimed their weapons through open windows of enemy-held buildings in the hope of killing the defenders inside. The success rate had been stunningly low. The rockets did not detonate until they hit something solid, and, when they did, the blast was in the direction the rocket was going—away from the NVA manning the windows. Late-night bull sessions resulted in new orders to all the rocketmen. On the morning of February 4, they fired their rockets around—rather than through—the windows from which the NVA were firing, counting on the blast to send masonry shards and shrapnel ricocheting around the room to cut down the men inside.[3]
>
> —Eric Hammel

---

[3]Hammel, *Fire in the Streets*, p. 155.

likely to possess the high profile of "big ticket" weapon systems that attract disproportionate attention during funding debates.

It was evident, therefore, that a significant majority of those attending the conference believed change is essential. Yet change itself holds no promise of success or improvement. As one conference participant explicitly stated, there is also the need to "get it right." What "right" is, and how to get there, is perhaps now somewhat more clear, thanks to the input of the nearly sixty men and women at this gathering. Far more work lies ahead before firm solutions are found, yet the time until the next commitment of U.S. forces into an urban environment is most likely but months. An enemy seeking asymmetric advantages will be hard pressed to find an alternative more likely to neutralize U.S. superiority than urban operations. With increases in military MOUT readiness, National Command Authorities can assign forces to stability or support missions in cities or send them into urban combat confident that national interests will be well served. Without these improvements, and without military leaders' and policymakers' recognition of the importance of MOUT readiness and the inevitability of future operations in urban environments, these national interests are at risk. It is essential that movement toward improving MOUT readiness begin as soon as possible.

Part 2
Appendixes

CONFERENCE AGENDA

RAND Arroyo Center

and

United States Army Dismounted Battlespace Battle Lab

**UNITED STATES MILITARY OPERATIONS ON URBANIZED
TERRAIN READINESS, 1998–2025**

24–25 February 1998

**RAND
1333 H Street, N.W., 8th Floor, Washington D.C.**

Tuesday, 24 February

0745   Check-in and Continental Breakfast

Conference Room 806

0815–0830   Welcome
Dr. Thomas L. McNaugher, Associate Director, RAND
Arroyo Center

0830–0915   Introductory Remarks
Dr. Fenner Milton, Office of the Assistant Secretary of the
Army (Research, Development and Acquisition)

0915–1000   The Russian MOUT Experience—Grozny
Mr. Timothy Thomas, Foreign Military Studies Office

Tuesday, 24 February (continued)

1000–1100   USAIS and ACTD MOUT Initiatives
LTC Randall Sullivan, MAJ Brad Sargant,
Dismounted Battlespace Battle Lab

1100–1115   **Break**

1115–1200   Entering the Intersection: Choosing the Right MOUT
Strategy for the Twenty-first Century
Dr. Russell W. Glenn, RAND Arroyo Center

1200–1530   **Working Lunch** *(Conference Rooms 802, 806, 1059, 1061)*

Identify major shortcomings in current U.S. military
MOUT readiness and propose a 1998–2005 strategy for
improvement.

Conference Room 806

1530–1600   First Group Briefing

1600–1605   Second Group Set-up

1605–1635   Second Group Briefing

1635–1700   Concluding Remarks and Comments for Second Day

1700   **Wine & Cheese Wind-down**

Wednesday, 25 February

0745   Continental Breakfast

Conference Room 806

0800–0845   MOUT in Bosnia
LTC Tony Cucolo, U.S. Army War College

0845–0930   Sea Dragon and Urban Warrior
LTC Gary Schenkel, USMC Warfighting Lab

0930–1000   **Break**

1000–1045   State of MOUT Simulations
Dr. Randall Steeb, RAND Arroyo Center

Wednesday, 25 February (continued)

1045–1130   MOUT in Vietnam—Hue, 1968
LTG George R. Christmas (USMC, Ret.)

1130–1400   **Working Lunch** *(Conference Rooms 802, 806, 1059, 1061)*

Develop a strategy for developing MOUT readiness in
1998–2025. Strategies for 1998–2005 developed on
Tuesday may be used, altered, or rejected completely.

Conference Room 806

1400–1440   First Group Briefing

1440–1445   Second Group Set-up

1445–1525   Second Group Briefing

1525–1600   Concluding Remarks

CONFERENCE ATTENDEES

The conference comprised 61 representatives from the following organizations: Office of the Assistant Secretary of the Army for Research & Technology (SARDA); United States Army Infantry School; United States Marine Corps Commandant's Warfighting Laboratory; J6 and J8 Joint Staff sections; Headquarters, Department of the Army Deputy Chief of Staff for Operations (DCSOPS); AMSAA; United States Army Joint Readiness Training Center; United States Marine Corps Studies and Analysis Division, MCCDC; United States Army Natick Research, Development and Engineering Center; Institute for Defense Analysis; Foreign Military Studies Office; United States Army Command and General Staff College; United States Army Soldier Systems Command; United States Marine Corps Warfighting Development Integration Division; Office, Deputy Under Secretary of Defense (Advanced Technology); United States Army Training and Doctrine Command (TRADOC); Pacific Northwest National Laboratory and Logistics Integration Agency; United States Army War College; Omega Training Group; Science Applications International Corporation; United States Army Deputy Chief of Staff for Intelligence (DCSINT); National Ground Intelligence Center; USARDSG (UK); United States Army TACOM-ARDEC; United States Marine Corps Strategy and Plans Division; United Kingdom's AM (Infantry Combat), CDA (Land); Headquarters, Department of the Army, Strategy, Plans, and Policy Division; United States Department of Energy; Applied Research Associates, Inc.; Booz-Allen & Hamilton Associates; United Kingdom's SOZA (LW) D, DOR (Light Weapons and Simulation); the United States retired military community; and the RAND Arroyo Center.

Those attending the conference:

Mr. Scott Bamonte

Ms. Kristen Baldwin

Mr. David S. Barnhart

Colonel Timothy G. Bosse

Major Michael A. Browder

Major Kevin Brown

Ms. Susan J. Butler

Mr. Chris Christenson

LTG George R. Christmas
(USMC, ret)

Mr. Raymond Cole

LTC Tony Cucolo

LTC Michael Dolby

Mr. Art Durante

Major Michael C. Edwards

Mr. Sean J. Edwards

Major (P) Robert E. Everson

Ms. Carol Fitzgerald

Major David R. Folsom

Colonel John Fricas

Mr. David Fuller

Mr. Scott Gerwehr

Dr. Russell W. Glenn

SFC Darrin P. Griffin

Mr. John Gordon

Mr. Rafael Gutierrez

Lieutenant Colonel Robert
F. Hahn

Dr. Kenneth Horn

Mr. Bob Jordan

Mr. James O. Kievit

Colonel James Lasswell

Lt Col Roger Lloyd-Williams

Mr. John Matsumura

Dr. Thomas L. McNaugher

Mr. Tom McNerney

Dr. Cynthia J. Melugin

Dr. A. Fenner Milton

Mr. Patrick C. Neary

Dr. Robert O'Connell

Dr. John Parmentola

Dr. Roy E. Reichenbach

Mr. William Rosenau

Major Robert J. Rush

LtCol. J. Tyler Ryberg

LTC Gary W. Schenkel

Mr. Neil Shepherd

Mr. Sam Spears

Lieutenant Colonel Robert Snyder

Dr. Randall Steeb

Brigadier General Robert
J. St. Onge, Jr.

Mr. Kevin Stull

Mr. George L. Suarez

LTC Randall Sullivan

Mr. Timothy L. Thomas

Mr. Milivoj Tratensek

Lieutenant Colonel Barry
N. Tyree

Mr. Robert B. Underwood III

Lieutenant Colonel Patrick Vye

Mr. Kevin Wainer

Mr. Jimmy L. Walters

Major Charles R. Webster

Major Gerhard Wheeler

Colonel Tony Wood

PRESENTATION BY DR. A. FENNER MILTON

A Little History
The Defense Science Board Task Force

"We cannot destroy or significantly damage the infrastructure of a foreign urban center in pursuit of mission attainment and expect the population to remain friendly to either U.S. forces or those we support."

November 1994 Report Findings:

- MOBA is, and will continue to be, a major area of concern of U.S. Forces
- No single technology (including nonlethal technology) has the potential to revolutionize this type of operation
- Dramatic improvement in the effectiveness of MOBA can be achieved by integrating existing and new technologies under appropriate operational doctrine developed explicitly for MOBA
- Technology already exists, or could be developed rapidly, that can fill important requirements in MOBA
- Improving MOBA requires a systems approach - Now, as before, implementation of improvements is the challenge

Recommendation: Establish an ACTD program for MOBA

DSB.ppt   20 Feb 98

The MOUT Challenge

How to Gain the Technological Advantage
in Complex Terrain?

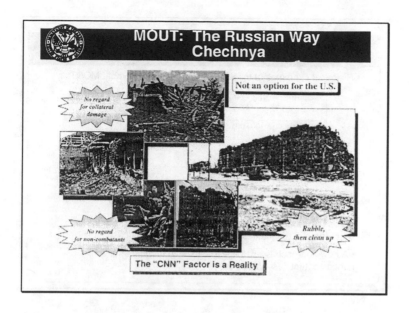

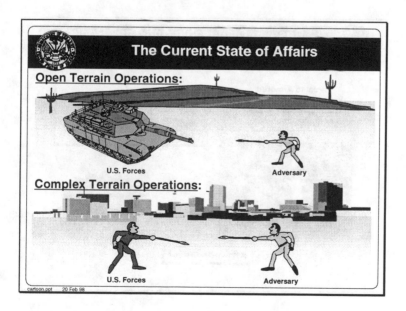

The Need for A New Approach

• Historical examples of militarily successful close combat MOUT:

Battle	Noncombatant KIA	US KIA	Noncom: Friendly KIA Ratio
Manila (1945)	100,000 (est)	1010	100:1
Hue (1968)	5,800	150 US + 400 ARVN	11:1
Panama (1989)	202	26	8:1
Mogadishu (1993)	500+	18	30:1

• Close Combat MOUT has resulted in mission accomplishment with **high** U.S. casualties and **high** noncombatant casualties

• A new approach is essential to minimize friendly and noncombatant casualties

ARROYO CENTER

Russian Disposable Anti-Tank Rocket Launcher

Threat to Armored Vehicles is Increased in Built-up Areas - Attack from Side and Rear

200 Meter Range

Asymmetric Threat

MOUT ACTD Objectives

- Improve the Near-Term Operational Capabilities of Soldiers and Marines through Experimentation with Technologies and the Development of Associated Tactics, Techniques and Procedures

- Develop Advanced Concepts for Army XXI Through to Army After Next to Guide S&T Investments

Military Operations in Urban Terrain (MOUT) ACTD

Objective: Through experimentation, demonstrate improved C4I, engagement, force protection, and mobility capabilities for conducting urban terrain operations.

Develops new tactics, techniques, and procedures

The most likely and resource intensive battlefield in the 21st Century

Integrates technology into a system of systems

Increased Operational Capabilities of Soldiers and Marines in MOUT

MOUTACTD.ppt   20 Feb 98

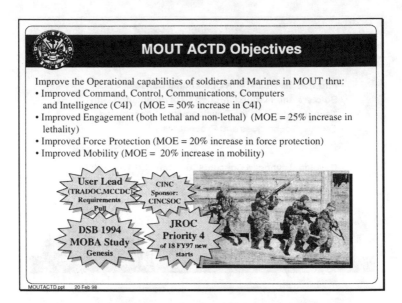

MOUT ACTD Objectives

Improve the Operational capabilities of soldiers and Marines in MOUT thru:
* Improved Command, Control, Communications, Computers
  and Intelligence (C4I)   (MOE = 50% increase in C4I)
* Improved Engagement (both lethal and non-lethal) (MOE = 25% increase in lethality)
* Improved Force Protection (MOE = 20% increase in force protection)
* Improved Mobility (MOE = 20% increase in mobility)

User Lead (TRADOC,MCCDC) Requirements Pull

CINC Sponsor: CINCSOC

DSB 1994 MOBA Study Genesis

JROC Priority 4 of 18 FY97 new starts

MOUTACTD.ppt   20 Feb 98

Capability Enhancements

Dismounted Force Connectivity and Situation Awareness	**Urban Land Warrior** — GPS Independent Position Navigation / Robust Communications
Lethality for Complex Terrain	**Fired from Enclosure** **Tank Killer** - e.g. Javelin **Bunker Busting** - e.g. Multi-Purpose Individual Munition **Defilade Kills** - e.g. Objective Individual Combat Weapon Objective Crew Served Weapon
Force Protection	**Unexposed Firing** (Helmet Mounted Display) **Anti Sniper Systems** **Booby Trap and Land Mine Detectors** **Individual Ballistic Protection**
Standoff Precision	**Enhanced Fiber Optic Guided Missile** **Precision Guided Mortar Munition**

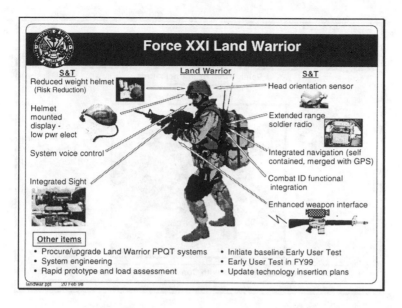

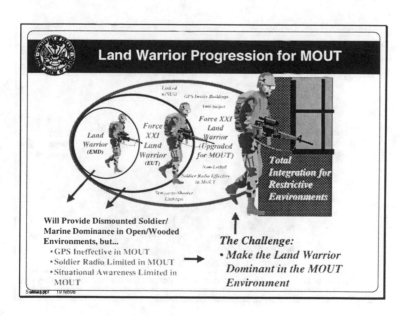

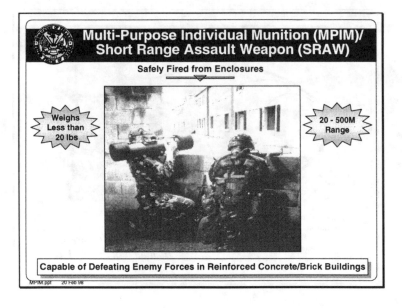

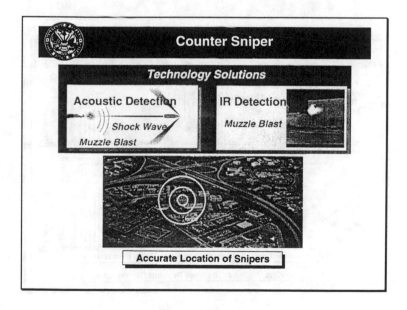

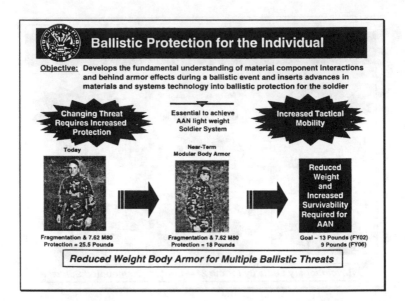

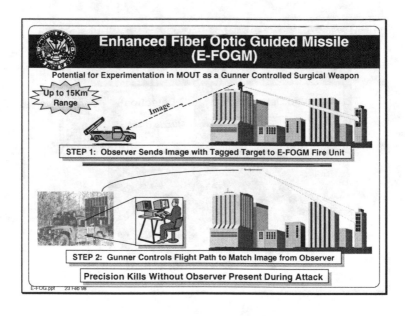

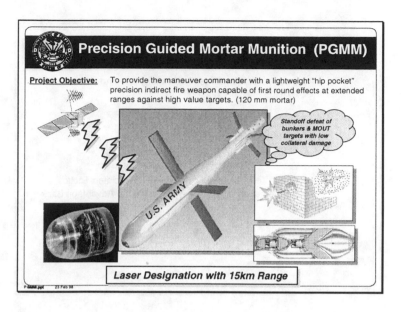

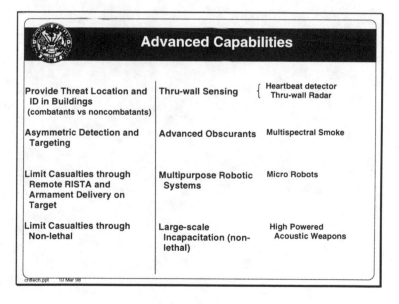

Summary

- Adversaries Utilizing Older Technology have Advantages in MOUT Due to the Increased:

  - Protection by Terrain
  - Cover (non-combatants)
  - Degradation of U.S. Technologies

- Advanced Technologies Coupled with Associated Tactics, Techniques and Procedures will Provide the Advantage Back to the U.S.

summary.ppt   20 Feb 98

PRESENTATION BY MR. TIMOTHY THOMAS

THE BATTLE FOR GROZNY
JANUARY, 1995

Lessons learned from and Russian
opinions about the most lethal
combat-in-cities experience since
Berlin (1945), Hue (1968), and
Beirut (1983)

CONFERENCE OBJECTIVES
(Russian Opinions)

- MOUT readiness: no country is ready
  - no standard, depends on size of city, type of resistance, corr of "other" forces, etc.
- MOUT debate: must continue to adjust
  - opn concepts: police action versus c-in-c
  - tng techniques: psychol, comm, assault tms
  - simulation requirements: clean versus dirty icon
- Recommended technologies: still changing
  - combatant ID, non-lethal wpns, etc.

FORCE XXI AND THE COMPUTER ICON

- A 21st century force can easily find Adid and win in Somalia!
- Even an outdated, untrained force can find President Dudayev in Grozny!
- U.S. thinks simulation centers save the day

BEHIND THE CLEAN ICON

- 4,000 rounds fell per hour in Grozny; 350 fell per hour in Sarajevo
- Russians lost 20 of 26 tanks and 102 of 120 BMPs in the initial battle
- Russian soldiers were hung upside down in windows or hung on crosses in city center
- By February 7 one-seventh of the Russian brigade had viral hepititus
- Police actions versus combat-in-cities

FORCES

- Russian
  - 38,000 men, 6,000 in the attack
  - 230 tanks
  - 454 BMPs
  - 388 tubes

- Chechen
  - 15,000 men
  - 50 tanks
  - 100 armored vehicles
  - 60 tubes
  - 150 anti-aircraft guns
- Plus the press, local population, knowledge of the city, and Allah

---

 # THE PLAN

- Russian
  - three pronged attack from the north, west, and east
  - left the south open
- Required high level of mvt and coord in dictated time frames with inadequate recon and comms with HQ

- Chechen
  - three concentric circles from the city center at distances of 1-1.5 km, 5 km, and the city outskirts
  - multiple ambushes
  - channelize
- Moved easily, planned destruction of refineries, chem plants

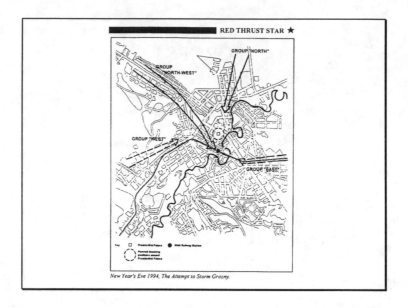

New Year's Eve 1994, The Attempt to Storm Grozny.

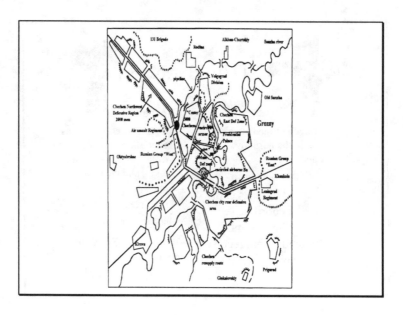

CHECHEN TACTICS

- Let armored columns into the city, seal off the city, methodical annihilation
- Vehicle kills of lead/rear vehicles
- Shoot from tops of buildings or from basements where guns couldn't reach

- Shoot for legs, then shoot those coming to help
- Booby trap doorways, breakthrough areas, entrances to metros and sewers, bodies

 # LESSONS

- From OODA to OCODA
- Turning the local population against you
- Identifying local combatants
- Psychological impact on the combatant requires a strong reserve force

- Deception, mines/traps
- Motorola, Sony TV stations, the Internet
- Snipers, grenade launchers, anti-tank weapons
- Three-dimensional fight-do weapons fit?
- Preparation, planning, discipline, execution

LESSONS LEARNED (cont.)

- Flame-throwers worked against strong points and snipers
- Boundaries between units still the weak spot; Russians on one floor, Chechens above on next floor

- Send out "contact ambushes" to find ambushes
- Use lots of infantry/marines
- Cellular comms versus FM
- Three tier ambushes
- Baiting/hugging
- Health of soldiers

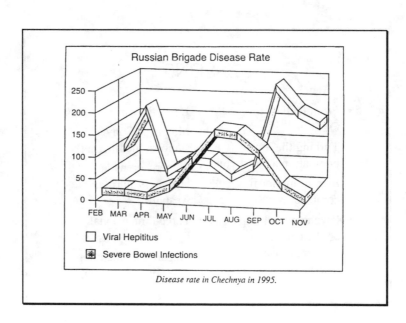

Disease rate in Chechnya in 1995.

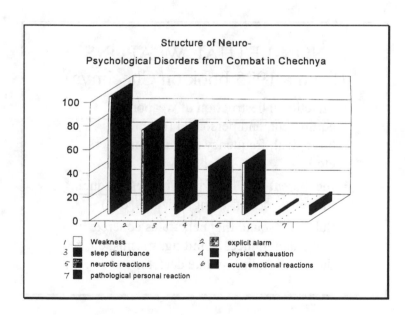

←Psycho-physiological support for infantry and airborne

- 65% of combat activity is personnel dependent, 35% a function of tech means
- 1312 troops surveyed
- 72% had some type of psychological disorder, 28% were healthy
- Need 5 clinical specialists at the army level, 2 psycho-physiologists, 1 psycho-pharmacologist, 1 psychiatrist, 1 medical psychologist

Structure of Neuro-Psychological Disorders from Combat in Chechnya

/	Weakness	2	explicit alarm
3	sleep disturbance	4	physical exhaustion
5	neurotic reactions	6	acute emotional reactions
7	pathological personal reaction		

going to Grozny through Bamut by his Chechen driver, a native of Lower Bamut? Cuny had most likely intend- knight of philanthropy and Texas adventure hunter, disappeared without a trace.

Army Hits Self-Destruct Button

The drawn-out conflict
in Chechnya has begun
to destroy its players – the troops.

By Alexander ZHILIN,

MN National Security Editor

Not being trained or prepared for prolonged policing and punitive operations, the army has entered a period of rapid disintegration, due mostly to never-ending military operations that lack any clear objectives.

Military suppliers generously provide deeply depressed Russian soldiers with vodka, while local residents see to it that they never run out of narcotics. Intoxication (by alcohol or drugs) is the standard state of the Russian soldier in Chechnya. This often results in tragedy as drunken quarrelling soldiers shoot it out among themselves. The statistics are appalling: for every Russian soldier killed by separatists, five are killed in accidents, brawls, orarmed conflicts with one another.

It's especially important to note the commercialization of military operations in Chechnya.   Officers of the ever more serious danger to society.

army and interior troops openly claim the existence of a "price list" for retreat from encirclement and protection of certain towns from federal mop up operations. The sums mentioned : $20,000 to $40,000. It's hard to believe, but rebels have been known to slip through triple blockades without any problems.

Psychiatrists report that soldiers in Chechnya have completely lost sight of the meaning and motives behind their actions, and are therefore breaking down mentally at an alarmingly rapid rate. Many approach a state of lunacy. Doctors believe that anyone who has fought in Chechnya for more than three months requires serious psychiatric treatment.

The Defense Ministry has a different view on this point: the army is acquiring combat experience. The majority of army personnel has served in Chechnya, and a significant number of soldiers is on a second or third tour.

As the Kremlin weighs the different possible solutions to the Chechen crisis in the interests of the presidential campaign, the army continues down a path of self-destruction, becoming an

---

NON-LETHAL WEAPONS
(from a 1995 book on Chechnya)

- Functional destruction of weapons, equipment, and personnel (chemical, biological, frequency modulation, lasers, etc.)
- Incoherent rays of light (blindness, reducing the sense of well-being, seizures, etc.)
- Subsonic sound that penetrates concrete or metal and induces vomiting, spasms, etc.-- the never ending smoke detector

NON-LETHAL WEAPONS
(cont.)

- Chemical and biological
  - "traction interrupters" to interfere with equipment's working parts
  - chemical paralyzers of people
  - pyrophoric materials to burn non-flammables
  - change road surfaces (slippery, etc.)
  - biologically destructive materials (destroy electricity and insulating materials)
- Holograms, aerosols, smoke

PRESENTATION BY LTC RANDALL SULLIVAN

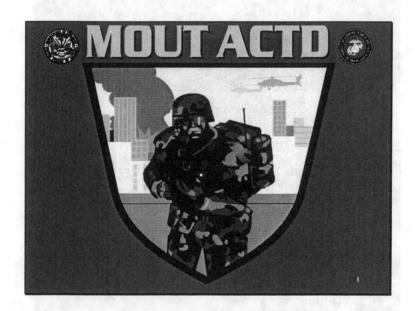

MOUT CONFERENCE

LTC RANDY SULLIVAN -- DBBL

MAJ BRAD SARGENT -- MCWL

MOUT ACTD

AGENDA

- INTRODUCTION
- MOUT THRU 2015
- MOUT ACTD

MOUT ACTD

MOUT THRU 2015

- CONCEPT BASED REQUIREMENTS SYSTEM
- TECHNOLOGY AS ENABLER
- TRAINING/EQUIPPING IS KEY
- COMBINED ARMS FIGHT
- CLOSE COMBAT <u>WILL NOT</u> GO AWAY
- REQUIREMENT FOR CONTROL DOES NOT GO AWAY

4

MOUT ACTD

CONTROL

- INFORMATION
- RESOURCES
- NON-COMBATANT ISSUES
- COMBAT BEHAVIORS

5

MOUT ACTD

- FOCUS ON BATTALION AND BELOW
- INCREASE CLOSE COMBAT/CLOSE FIGHT
  CAPABILITY
  - C4I
  - SURVIVABILITY
  - LETHALITY
  - MOBILITY
- THERE IS A USE FOR INDIRECT FIRES
  (SURFACE AND AIR)
- REDUCE BUT NOT ELIMINATE NON-
  COMBATANT DEATHS

6

MOUT ACTD

CAUTIONS

- SOLUTION SETS MUST BE IN "ART" OF POSSIBLE
- SOLUTIONS MUST BE ACCEPTED BY
  OPERATIONAL COMMANDER
- SOLUTIONS MUST COMPLY WITH TREATIES:
  - SLEEPER AGENT
  - DIRECTED ENERGY (LASERS)
  - EMP
- NON-LETHAL HAS BECOME A "CATCH-ALL"
- CASUALTY-FREE COMBAT IS A "MYTH"
- CURRENT TRAINING FACILITIES ARE
  INADEQUATE

7

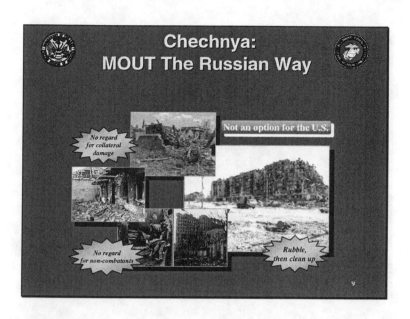

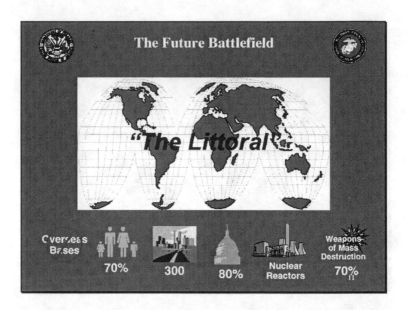

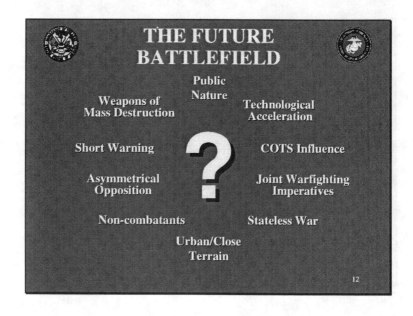

URBAN OPERATIONS

- AN ENVIRONMENT, NOT A MISSION
- MANPOWER INTENSIVE
- SHORT RANGE ENGAGEMENTS
  "Gunfight"
- TECHNICAL ADVANTAGE OF U.S.
  FORCES GREATLY REDUCED

13

Urban Warfare Tactics
Urban BattleSpace

- Confined , Linear

- High density of non-combatants

- Man-made features leverage control

- "Nodal" Warfare

14

Urban Warfare Tactics

Traditional Urban Warfare

- Surround and isolate city

- Conduct linear, methodical sweep to clear enemy

- High consumption rates for ammunition

- High casualty rates

15

**REQUIREMENTS OF MOUT
"JOINT PERSPECTIVE"**

- REVOLUTIONARY WAR
- WAR OF 1812
- CREEK INDIAN CAMPAIGN
- SEMINOLE INDIAN CAMPAIGN
- MEXICAN WAR
- HARPER'S FERRY
- AMERICAN CIVIL WAR
- RAILROAD STRIKE OF 1877
- BOXER REBELLION
- PHILIPPINE INSURRECTION
- MEXICAN EXPEDITION

- CHINA SERVICE
- WORLD WAR I
- WORLD WAR II
- KOREAN WAR
- VIETNAM
- GRENADA
- THE GULF WAR
- HAITI
- SOMALIA
- BOSNIA

16

REQUIREMENTS

- SERVICES AND CINC INPUTS FOCUSED THE MOUT ACTD ON FOUR PRIMARY AREAS
- ARMY & MARINE REQUIREMENTS SHARE COMMONALITY
- "ROOM CLEARING" IS THE SAME FOR THE SOLDIER OR MARINE

17

1. COMMAND, CONTROL, COMMUNICATIONS, COMPUTERS AND INTELLIGENCE (C4I)

- POSITION LOCATION
- COMMON OPERATING PICTURE
- NON-LOS COMM.

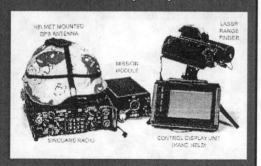

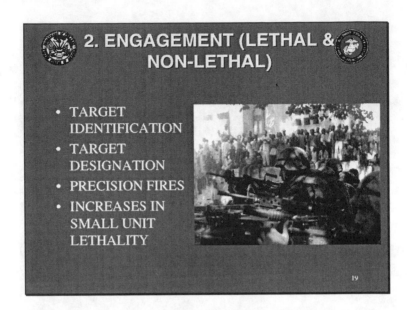

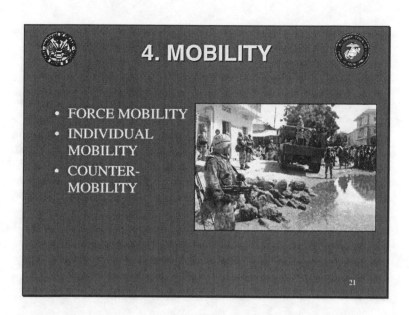

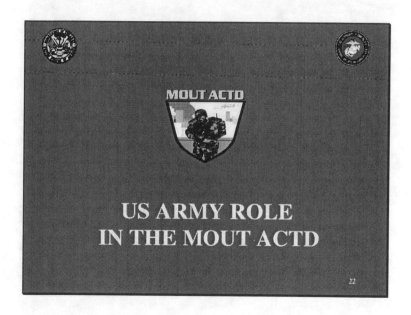

MOUT ACTD

- PROGRAM LEAD
- OPERATIONAL CO-MANAGER
- EXPERIMENTS CONDUCTED BY DISMOUNTED BATTLESPACE BATTLE LABORATORY
- OPERATING/EXPERIMENTAL FORCES PROVIDED BY XVIII ABN CORPS

23

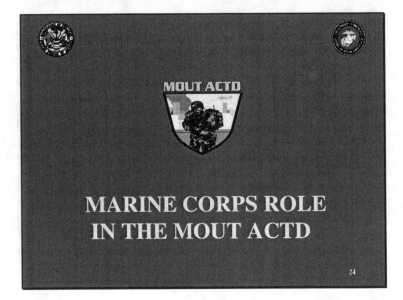

MARINE CORPS ROLE IN THE MOUT ACTD

24

MOUT ACTD

- USMC OPERATIONAL CO-MANAGER
- EXPERIMENTS CONDUCTED BY
  MARINE CORPS WARFIGHTING
  LABORATORY
- OPERATING/EXPERIMENTAL FORCES
  PROVIDED BY II MEF
- AN IMPORTANT PART OF URBAN
  WARRIOR

25

MOUT ACTD

JOINT EXPERIMENTS

- FT BENNING & CAMP LEJEUNE
- EVALUATES SYNERGY OF ALL
  SELECTED TECHNOLOGIES & TTP'S
- LATE 1999

26

MOUT ACTD

CULMINATING DEMO

- JRTC FT POLK
- US ARMY INF. BN WITH USMC RIFLE COMPANY (REIN) ATTACHED
- SEPTEMBER 2000

27

OTHER JOINT INITIATIVES

- SMALL UNIT OPERATIONS (SUO)

28

SMALL UNIT OPERATIONS

- DARPA PROGRAM
- DISMOUNTED BATTLESPACE BATTLE LABORATORY AND MARINE CORPS WARFIGHTING LABORATORY ACTIVE PARTICIPANTS
- DISMOUNTED BATTLSPACE BATTLE LAB IS EXPERIMENT LEAD
- FOCUS IS ON
  - C4I TECHNOLOGY / SENSORS
  - ROBOTICS
  - SITUATIONAL AWARENESS

29

SMALL UNIT OPERATIONS

- GOAL = "GIVE SMALL UNITS CLOUT ON RESTRICTIVE TERRAIN"
- SUPPORTING PROGRAMS:
  - ADVANCED FIRE SUPPORT SYSTEM "MISSILE IN A BOX"
  - MICRO UAVs
  - DISTRIBUTED COMMAND POSTS
  - TACTICAL COMMUNICATIONS

30

PRESENTATION BY DR. RUSSELL W. GLENN

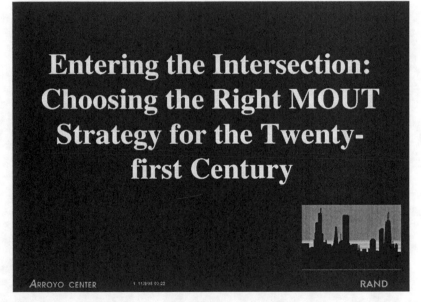

This presentation is part of RAND Arroyo Center's ongoing research on military operations on urbanized terrain (MOUT) for the Office of the Secretary of the Army for Research, Development and Acquisition (SARDA) and the United States Army Deputy Chief of Staff for Operations (DCSOPS). Previous Arroyo Center efforts in this area include:

Marching Under Darkening Skies: The American Military and the Impending Urban Operations Threat, Russell W. Glenn, MR-1007-A, 1998.

"Military Operations on Urban Terrain: An Invitation to Warfare in the 21st Century," Randy Steeb, John Matsumura, Russell Glenn, and Jon Grossman, unpublished RAND Arroyo Center research, September 1997.

Combat in Hell: A Consideration of Constrained Urban Warfare, Russell W. Glenn, MR-780-A/DARPA, 1996.

Objective

Develop a MOUT strategic concept for
the period 1998-2025 and the ACTD-
>Force XXI->AAN transition

The objective of the briefing is to review initial work on the development of a strategic concept with application during the periods envisioned for the MOUT Advanced Concepts and Technology Demonstration (ACTD), Force XXI, Army After Next (AAN), and beyond.

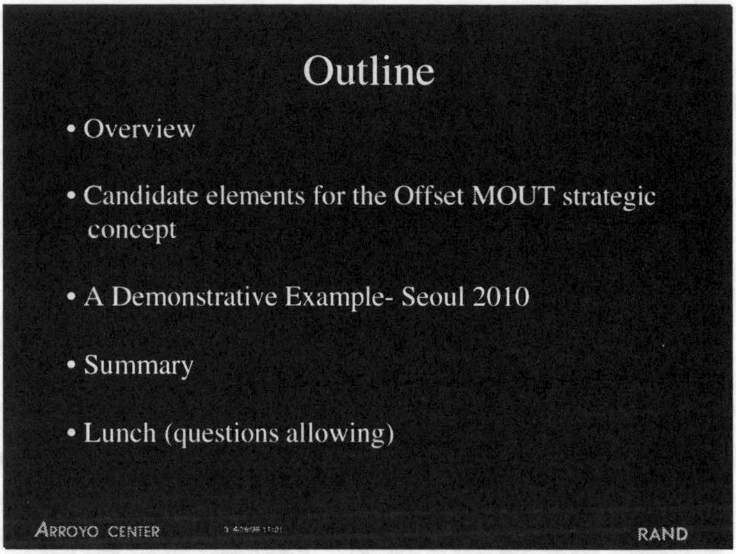
Outline
• Overview
• Candidate elements for the Offset MOUT strategic concept
• A Demonstrative Example- Seoul 2010
• Summary
• Lunch (questions allowing)
ARROYO CENTER
RAND

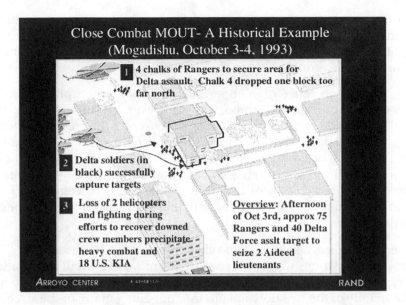

The costs of close combat MOUT have been definitively and repeat-edly demonstrated in modern military operations. Manila, Aachen, Stalingrad, Berlin, Seoul, Hue, Suez City, Khorramshahr, Panama City, Grozny, and Mogadishu are but a sample. The last, America's most recent excursion into urban warfare, helps us to understand the challenges MOUT will present to U.S. forces in the future. By the time operations ended less than 24 hours after their initiation, eigh-teen Americans had been killed. Another 73 were wounded. The vast majority of these 91 casualties were men in the original approx-imately 145 infantrymen and aircraft crew members who partici-pated in the initial assault to capture two of Mohammed Aideed's subordinates. These losses represented a casualty rate of over 60%. In excess of 500 Somalis, many of them noncombatants, were also killed.

Soldiers rappelling into the target area just shy of 4:00 PM on October 3rd immediately received incoming fire. Somali sentiments, encour-aged by local clan leaders, were largely anti-U.S., having been far more influenced by the recent deaths of Somali leaders during an attack helicopter engagement than by the preceding months of United Nations and American humanitarian support. Several of the assaulting soldiers were pinned down by the intensity of the fire.

Movement was severely hindered by having to move wounded comrades. In the confusion, shock, and desperateness of street fighting, intentions to distinguish between armed Somalis and innocent civilians were often discarded.

The original objective of the operation was soon accomplished with the capture of Aideed's supporters. However, the mission quickly took on the added tasks of finding and recovering the crew members of two helicopters shot down by Somali rocket propelled grenade fire. These tasks, complicated by the difficulty of extracting the body of a pilot from one aircraft's twisted frame and the heavy fire received by wheeled vehicles attempting to reach the beleaguered U.S. soldiers, would carry the mission into darkness and through the next day's sunrise.

Enemy fires struck soldiers and vehicles from multiple directions. To move in the streets was to receive the attention of Somali automatic weapons. There was little respite from incoming rounds anywhere in the vicinity of the two downed aircraft, nor along many of the streets between the UN compound and target areas. Occupants of thin-skinned U.S. vehicles suffered during movement; their ultimate defense was to fire into any potential enemy hiding place. Three helicopters, in addition to the two shot down, were severely damaged while providing support.

Rules of engagement and lack of theater assets precluded virtually any "offset" support. Artillery, naval gunfire, and other indirect fires could not have been employed even if available due to the threat they posed to noncombatants and friendly forces engaging the enemy at very short ranges. There was no AC-130 to call upon. Protective vests offered some, but by no means completely effective, shielding. Superior American skills, discipline, and equipment were in large part neutralized by the short ranges, lack of intelligence, delayed command and control, inability to distinguish between enemy and noncombatant, and other characteristics of Mogadishu combat.

Candidate Elements for a Future MOUT
Strategy

Decisive elements:
• Close Combat
• Offset operations

Facilitating elements:
• Sector and Seal
• Nodal operations
• Noncombatant Control

ARROYO CENTER   RAND

The military leader confronting contemporary MOUT is therefore faced with a three-faceted conundrum: (1) the immediate need to keep his soldiers and marines out of urban areas, (2) a very high probability that military operations in cities will be essential, and (3) a current lack of capabilities to fight in such environments without taking and inflicting politically unacceptable casualties. It is necessary to improve the ability of U.S. forces to conduct close combat MOUT while moving toward an offset concept as quickly as technological, doctrinal, and training innovations will allow.

A RAND Arroyo Center strategic concept created with these requirements in mind has the five elements shown. Two are potentially decisive in their influence, i.e., their application can result in successful accomplishment of a military mission involving urban combat. The three additional elements are facilitating; while not decisive, they have application in the service of mission accomplishment.

Though combat operations dominate discussion in this presentation, MOUT inherently includes all military operations in urbanized terrain, whether or not those operations involve combat. The five elements proposed here have application over the entire spectrum of urban operations. Our definitions of those elements are:

Close combat capabilities upgrade: MOUT operations involving short-range engagements between members of opposing forces or face-to-face interaction between friendly forces and noncombatants. Most engagements will occur within or in close proximity to buildings. The first element of RAND Arroyo Center's proposed strategic concept is an improvement in U.S. forces' capabilities to conduct close combat MOUT via adoption and adaptation of existent doctrinal, training, and technological resources.

Offset operations: Operations in which friendly forces engage the enemy or interact with noncombatants from remote locations. The offset acts to reduce the exposure of friendly force soldiers to enemy action or other threats. Engaging from offset positions requires accurate, timely intelligence and precision munitions.

Sector and Seal: Operations and activities with the primary purpose of (1) isolating selected groups or areas, or (2) denying access by selected groups to specified areas. Isolation may require use of devices (e.g., foams, smoke, etc.) to deny enemy reconnaissance, movement, or weapons use.

Nodal operations: Operations with the primary purpose of controlling key or decisive terrain, activities, or personnel in support of friendly force objectives. Examples are: (1) the seizure of commercial radio stations to deny enemy use of same while facilitating U.S. communications with members of the local populace, (2) denying an adversary access to water sources in order to preclude that enemy's coercive use of same, and (3) destruction of enemy C2 centers.

It should be noted that nodal operations in theory have decisive potential. For example, neutralization of a particular water source or elimination of a key leader could itself precipitate an enemy's surrender. However, instances in which such events have occurred generally included other actions critical to the downfall of a force. It is far more likely that nodal operations act in support of decisive close combat or offset operations than play a decisive role themselves.

Noncombatant control: Activities intended to influence noncombatant attitudes or behavior in a manner beneficial to friendly force efforts. Such activities could include:

1. psychological warfare efforts with the goal of obtaining noncombatant support for friendly force operations,

2. informing civilians of locations where food, water, or medical care are available, and

3. deception efforts undertaken to move noncombatants away from potential combat zones.

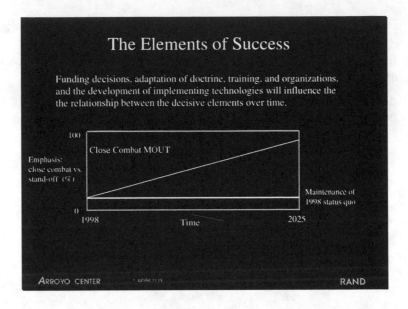

An alternative to close combat MOUT is essential if the armed services are to meet the new standards for operational success. While
enhancements of close combat capabilities are essential to ensure
military mission accomplishment at least cost in the short term, the
inevitability of MOUT in the future, and its similarly inevitable high
cost in casualties, requires transition to an offset urban operations
doctrine as rapidly as is feasible. The length of this transition will be
influenced by multiple factors; knowing that offset operations are the
objective will prepare military leaders so that they may more readily
seize opportunities to make advances in this regard.

A MOUT Strategy for 1998-2025:
Offset Urban Operations

• Three Phases:

  • Phase 1: Immediately upgrade current MOUT readiness
  • Phase 2: Capitalize on opportunities to introduce offset MOUT
  capabilities
  • Phase 3: Complete the transition to a predominantly offset
  MOUT doctrine.

- Doctrine
- Training
- Technology

• Solutions will require a comprehensive approach

• Greater active and reserve force commitment in American cities

• Tension will result between economy and need to develop and
purchase sufficient quantities for near-term operations

• Many of the techniques and technologies resulting from this change
will likely have application to operations in other environments.

ARROYO CENTER   RAND

The Arroyo Center MOUT strategy recognizes that the transition to an offset approach must await development of supporting doctrine, training, technologies, and, possibly, alternative organizations. Yet improvements in urban operations capabilities cannot wait for these developments. Immediate enhancement of current close combat capabilities is needed. However, recognizing that an offset strategy holds promise helps to establish the awareness essential to seizing opportunities necessary for a timely transition to that strategy.

Regardless of the phase, developments must be comprehensive in the sense that they include consideration of:

1. both operations involving combat and those that do not,

2. all three levels of operations,

3. recognition that future operations are likely to be significantly constrained by stated and implied requirements to minimize friendly casualties, noncombatant losses, and infrastructure damage,

4. operations in all types of urban environments, to include Third World shantytowns.

5. recognition that today's urban areas are of a size unknown in earlier years. The numbers of inhabitants, structures, and vehicles, and the sheer geographic magnitude of cities and their surrounding urban sprawl, have increased dramatically. Seizing and/or controlling complete metropolitan areas may be beyond the capabilities of any contemporary military force. Alternatives to these end states, e.g., moving from a secure base to an objective and returning, need development.

6. both offensive and defensive operations during MOUT missions entailing combat.

Enhancements also need to recognize that MOUT is not an international contingency alone. The 1992 Los Angeles riots demonstrated that both reserve and active duty forces may well be involved in domestic urban operations.

The tension between improving close combat MOUT and transitions to an offset approach will demand difficult funding decisions. One factor influencing these decisions is the applicability of many changes to doctrine, training, structures, and technologies to operations in other environments.

MOUT Doctrinal Issues

Written doctrine status:
- Doctrine
- Training
- Technology

- US Army's FM 90-10-1 a step in right direction
- Best combined arms MOUT manual in U.S. military is MCWP 3-35.3
- FM 90-10 scheduled for rewriting
- Few subordinate manuals are up to date with respect to MOUT

Doctrinal issues requiring greater attention:

- MOUT as more than combat operations
- Virtually no operational level MOUT doctrine.
- Need doctrine for other than securing entire built-up areas
- Widening of capabilities gap between wealthy and other nations
- Noncombatant considerations demand more coverage
- More coverage of regular-SOF and USAF MOUT operations needed
- Changed areas of urban operations. i.e.. no longer predominantly Western Europe; shanty towns

ARROYO CENTER   RAND

MOUT doctrine has improved in the last decade, but the enhancements have come too slowly and reflect many of the shortcomings highlighted in the notes to the previous slide. The Army's Field Manual 90-10-1, *An Infantryman's Guide to Combat in Built-up Areas*, advanced beyond the outdated 1979 edition of FM 90-10, *Military Operations on Urbanized Terrain (MOUT)*, with its introduction of two new concepts: surgical and precision MOUT. The recent Marine Corps Warfighting Publication 3-35.3 is better yet and could serve as a combined arms MOUT manual of value for all services pending publication of a truly joint manual.

However, both of these manuals lack the comprehensive scope necessary for today's strategic environment. More attention to elements of MOUT other than combat, political intolerance of casualties, levels of operations above the tactical, the changing character of modern urban areas, and domestic operations is necessary. A recent decision to update FM 90-10 is encouraging; it is also an opportunity to redress these shortfalls.

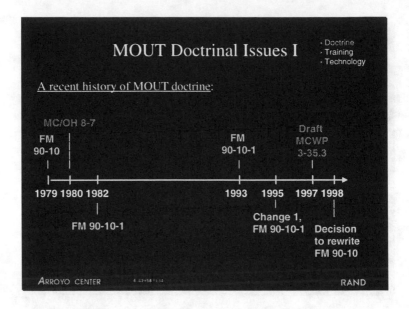

Updates to MOUT doctrine have been a long time in coming. World War II-style doctrine as presented in the late 1970s and early 1980s only recently has undergone even moderate change. The renewed interest is encouraging, but even these new manuals fall short of today's urban operations requirements.

MOUT Doctrinal Issues II

· Doctrine
· Training
· Technology

Doctrinal issues and current manuals:

Doctrinal Issue	FM 90-10	FM 90-10-1	MCWP 3-35.3
Combined arms	POOR	FAIR	GOOD
Weapons effects	POOR	FAIR	GOOD
Noncombatant considerations	POOR	FAIR	FAIR
Regular-SOF considerations	POOR	FAIR	FAIR
Third World structure types	POOR	POOR	FAIR
Joint Operations	POOR	POOR	FAIR
Multinational/Interagency	POOR	POOR	POOR
More than combat operations	POOR	POOR	POOR
Operational level MOUT	POOR	POOR	POOR
Tasks other than secure entire city	POOR	POOR	POOR

where: POOR = Little or no mention of topic
FAIR = Some mention of topic with brief, but inadequate, discussion
GOOD = Considerable discussion, adequate or nearly adequate
...and then there are domestic MOUT concerns

ARROYO CENTER   RAND

MOUT doctrine has improved in the 1990s, but the enhancements have come too slowly and reflect many of the shortcomings highlighted in the notes to the slide before last. The Army's Field Manual 90-10-1, *An Infantryman's Guide to Combat in Built-up Areas*, advanced beyond the outdated 1979 edition of FM 90-10, *Military Operations on Urbanized Terrain (MOUT)*, with its introduction of two new concepts: surgical and precision MOUT. The recent Marine Corps Warfighting Publication 3-35.3 is better yet and could serve as a combined arms MOUT manual of value for all services pending publication of a truly joint manual.

However, both of these manuals lack the comprehensive scope necessary for today's strategic environment. A partial list of MOUT concerns requiring more coverage includes urban operations other than those involving combat, noncombatant considerations, operations above the tactical level, the changing character of modern urban areas, and domestic operations. A recent decision to update FM 90-10 is encouraging; it is also an opportunity to redress these shortfalls.

Other challenges are inherent in MOUT to come. The gap between U.S. operational and technological capabilities and those of coalition and alliance members is widening; this is likely to continue. Future missions are also likely to demand regular force–SOF cooperation in may scenarios, yet MOUT doctrine provides little guidance for such contingencies. USAF support, so critical in other environments, could and should be an integral part of urban scenarios as well; the difficulties posed by buildings, short ranges, target acquisition, munitions flight profiles, and other factors have yet to be overcome.

MOUT Training

- Doctrine
- Training
- Technology

• Lack of U.S. Army MOUT training sites + low priority on urban operations training = limited proficiency

• Some short-term training possibilities:

  • Staff rides and terrain walks in local built-up areas
  • Use abandoned buildings on installations
  • Develop viable suite of MOUT simulations
  • Create MOUT live-fire ranges and tables

• Longer term training can benefit considerably from use of virtual sand table, Big Cities system, and similar tools

ARROYO CENTER

RAND

U.S. military force readiness to take on MOUT missions is hindered both by a lack of training facilities and a sometimes lesser emphasis on urban operations than is appropriate for the threat level. Commanders have overcome fuel, funding, and other shortages in the past to maintain and improve capabilities in other areas; they must do the same in preparing for coming demands in the world's cities. In the longer term, simulations and decisionmaking aids will assist leaders in preparing units and staffs for these events.

Offset MOUT
Critical Technologies Demonstration

· Doctrine
· Training
· Technology

Concept	Demonstrative Technology
• Remote Acquisition and Tracking/Monitoring	AeroVironment Black Widow
• Thru-wall Sensing	Raytheon thru-wall radar
• Advanced Obscurants	NICO, Buck Multispectral Smoke
• Surgical Weapons	Raytheon EFOG-M
• Multipurpose Robotic Systems	RST MDARS-E
• Large-scale Incapacitation	SARA Acoustic Weapon
• Urban Common Picture	DIA Big Cities

ARROYO CENTER   RAND

The solutions to MOUT—finding ways to accomplish the military mission while minimizing friendly casualties, noncombatant losses, and infrastructure damage—will not be easy to discover or to implement. MOUT is a complex problem; solutions will be multifaceted and likely system-of-systems in character. For demonstrative purposes, seven concept areas are identified and discussions of one or two technologies for each are provided in the following pages.

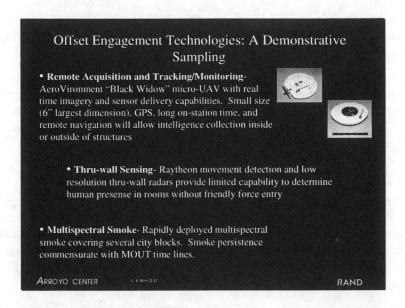

Remote Acquisition and Tracking/Monitoring: Both accomplishment of the military mission and casualty reduction will demand excellent intelligence acquisition and situation monitoring. UGVs, UAVs, biosensors, overhead photography, HUMINT, and COMINT are only a sampling of the systems and resources that will be employed to meet these requirements. Collection will have to include building interiors so as to pinpoint targets and allow for determination of an appropriate means of engagement. Intelligence tasks include locating potential targets, plotting access to targets, determining noncombatant locations, monitoring status, and collecting information in support of optimum engagement system selection.

An area of considerable promise is micro-UAVs. Aircraft of 18" (greatest dimension) and larger have demonstrated the capability to navigate to a specified location and return real-time video images. Current technologies involving systems of 6" size have been flown successfully, but work is ongoing to attain camera-carrying capability and non-line-of-sight navigation/image reception. Fixed-wing aircraft require considerable speed to remain aloft (20–25 mph); for MOUT purposes, therefore, rotary-wing or ornithopter (flapping wing) technologies appear to hold the greatest promise. Experiments using these latter approaches are in their early stages; functional prototypes could be available within the next five years. AeroVironment has a DARPA contract to continue work, but funding limits currently restrict work to fixed-wing prototypes.

Potential uses for micro-UAV systems go beyond intelligence collection. AeroVironment personnel see sensor delivery, signal relay, and lethal munitions delivery as viable future functions for these aircraft.

Thru-wall sensing: The enemy often prepares defensive positions in rooms and trains his weapons on the only access points. Attackers compensate by either preceding their entry with the throwing of a grenade and small arms fire, or by "stacking" soldiers and bursting in to engage an adversary as quickly as possible. The first technique presents dangers to friendly forces (fragmentation through walls, grenades bouncing back into friendly locations) and noncombatants that may be occupying buildings during clearing operations. The second has proven to result in high friendly casualties as enemy fire or booby traps take their toll on the attackers.

Soldiers need a way to "see" into rooms. Such a capability would speed unit movement through a structure as empty rooms could be marked, sealed, and bypassed. It would reduce both friendly and noncombatant casualties as enemy locations and armament could be identified before entering. Refinement of existent ammunition could allow linkage of this see-through capability with munitions capable of wall penetration so as to further reduce friendly force exposure via offset engagement.

Current radar technology offers limited resolution and penetration capability. Some models can detect both stationary and moving targets; others provide an audible tone upon detection of movement and have no stationary imaging capability. Future developments in thru-wall radar could provide greater detail on location, identification of the detection as enemy or noncombatant, and number of occupants in the room. Radar could also be used to detect booby-traps or other threats. Available systems provide crude imagery and as of yet lack the resolution necessary to accomplish these tasks.

Multispectral Smoke: In the near term, MOUT will require U.S. and coalition forces to conduct close combat operations. Multispectral smoke tailored to defeat an adversary's vision enhancement systems has the potential to obscure enemy detection and acquisition while permitting friendly forces to "see" via "windows" that do not inter-fere with friendly force viewers or other navigation aids. Such aids could include narrow bank laser radars, far infrared sensors, en-hanced UV imaging systems, and imaging MMW radars.

Multispectral smoke that blocks visible through LWIR radiation has been developed. Smoke capable of blocking 35 and 70 GHZ is also available. To date no one has specifically developed smoke with spectral windows. This is apparently due to a lack of funding for this concept. It is possible that current smoke products may have win-dows in the UV or in the far IR (>14 microns), but no one has ana-lyzed smoke in these spectral regions.

Even if no windows exist, inertial navigation devices, MMW radars, force protection systems, and other tactical adjustments could facili-tate effective close combat MOUT operations at reduced cost through smoke usage. JANUS modeling could help determine the feasibility of this approach.

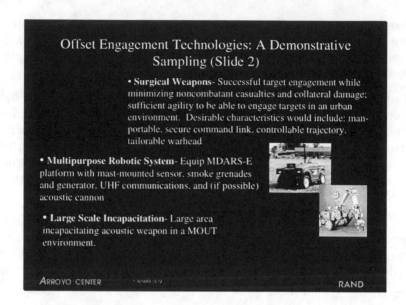

Surgical Weapons: Few Army systems possess the precision characteristics essential to successful engagements in cities. Narrow streets, sharp turns, and the close proximity of other structures result in high collateral damage and significant risk of noncombatant losses when area weapons are employed. With limited exceptions, strict ROE translate to minimal fire support for soldiers in close combat.

Most advanced indirect-fire weapons have trajectories that pose critical limitations on urban environment use. Generally, these limitations are directly associated with the complexity of engaging a specified target in an urban setting. Two key reasons are: (1) targets are frequently blocked by buildings or other obstacles, and (2) weapons may not be capable of vertical plane attack, e.g., through a window.

The enhanced fiber optically guided missile (EFOG-M) shows much promise in efforts to address these problems. Compared to most indirect-fire weapons, the EFOG-M has a much greater degree of maneuverability and, in theory, can fly around buildings and engage remotely located targets in a vertical plane. Practically, however, initial analysis suggests that the EFOG-M's large size and limited ability to sustain high-G turns may limit its overall applicability. A smaller, more maneuverable missile (larger platform area per unit weight) with a variable thrust gas turbine propulsion system holds promise.

Multipurpose robotic system: One of the primary means of reducing casualties is use of robotic systems for more dangerous missions requiring friendly force exposure. Roles are evident for both macro- and micro-sized unmanned ground vehicles: forward reconnaissance, mine and obstacle clearing, dispensing smoke and foam (both of which may be noxious to humans), carrying and using nonlethal weapons (acoustic cannons, microwave devices), evacuating wounded, and resupplying forces.

UGVs must have speed, agility, protection, and even some autonomous control capability. As they are used under conditions of great exposure, they must be survivable. Speed, armor protection, stealth, and active protection systems could all play a part in enhancing survivability. The robots should to be able to seek cover when necessary and may lose their command link at times. The latter would necessitate an autonomous route-finding capability for mission completion.

Several near-term demonstrations are possible. Small systems such as the Andros six or eight wheeled/tracked system can carry small payloads and even enter buildings when necessary. A more dramatic demonstration could use RST's MDARS-E. This larger system can be equipped with a variety of sensors, has reasonable speed, and can carry up to a 500-pound payload. One test scenario could require the vehicle to provide itself with smoke concealment, navigate through the smoke with MMW radar or long-wavelength IR imager, move to an occupied building, and employ an onboard acoustic cannon to drive out enemy defenders.

Constructive and virtual simulation could extend such a demonstration by showing how the system would work in a combined arms urban mission using different assumptions regarding onboard equipment, environmental conditions, and threat responses.

Large-scale incapacitation: Many MOUT missions will take place in areas where combatants and noncombatants are intermingled. Rules of engagement will limit friendly force capabilities in these circumstances. Any broad-area, nonlethal weapon would be very useful. Ideally the weapon would significantly reduce the enemy's combat effectiveness. Noncombatants would be affected but not permanently harmed.

Recent developments in acoustic weapons by SARA and other companies have led to successful limited demonstrations of this technology. The technology appears to be inexpensive, simple, and rugged. Field testing is scheduled for mid- to late 1998. Incapacitating and lethal ranges need to be determined in addition to the duration of engagement required. Simulations can also be used to determine these weapons' capabilities. Given the sensitivities regarding system employment, it is expected that significant numerical modeling will be required (as opposed to field testing) in the development of this technology.

The following fictional scenario aids in demonstrating potential applications for previously identified technologies. Potential uses of the capabilities would depend on actual contingencies undertaken and development of these or other technologies in the interim years available for further development.

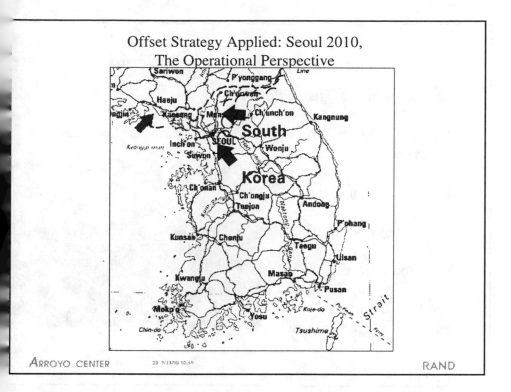

Offset Strategy Applied: Seoul 2010,
The Operational Perspective

ARROYO CENTER   RAND

In January 2008, North Korean armed forces attacked across the DMZ with their main effort focused on the seizure of Seoul. Advancing forces were halted after reaching the Han River's northern bank in the capital's metropolitan area. Allied plans call for a turning movement by two Republic of Korea Army (ROKA) corps with the objective of causing the North Koreans to withdraw from the urban area. ROK marines will complement the main effort with a demonstration from the Sea of Japan. A multinational ROK-US force is to attack from the south to clear Seoul.

The Tactical Situation in Seoul

- Movement and navigation difficult

- Republic of Korea Army (ROKA) communications hampered by
- urban conditions

- ROKA-US technology
- differential may cause
- capabilities disconnect

- Most South Korean civilians
- still in city, thus ROKA
- and US rules of
- engagement are very strict

ARROYO CENTER   25 8/18/98 10.59   RAND

Though allied forces control the southern banks of the Han, fifth columnists and the density of structures complicate preparations for the metropolitan area to the north. Twisting streets and a heterogeneous mix of modern and older architecture make navigation and selection of appropriate munitions very difficult. ROK communications are severely hampered by the terrain. Operations are further constrained by the large number of civilians remaining in Seoul and surrounding smaller cities on both sides of the river. Though many refugees fled south during the initial attack (blocking roads and delaying friendly forces attempting to move northward), the speed of North Korean movement precluded most from escaping their homes. ROK-US forces are operating under strict rules of engagement (ROE) to minimize ROK civilian casualties and minimize damage to the nation's infrastructure.

Preparation and Planning for the Attack

- Tall apartments and high ground
- provide dominating terrain for
- defenders
- Likely causes of culmination:
- manpower shortages and exhaustion
- Reconnaissance crucial to success
- ROKA/US commanders will seek to isolate battlefield at
- operational and tactical levels
- Concerns during the attack include friendly force
- concealment, interdiction of enemy reserve forces, and
- separation of enemy soldiers and noncombatants

ARROYO CENTER   27 9/18/98 10:59   RAND

Though the turning movement was successful in part, North Korean commanders left what appears to be a motorized division in the city, presumably to delay allied units attacking from the south. With all bridges across the Han damaged or destroyed, friendly forces faced an enemy positioned on dominating terrain, a mission demanding the crossing of a major river, and ROE that preclude extensive use of other than precision supporting fires were they to attempt to cross the water obstacle within Seoul. ROKA/US commanders therefore decide to cross the Han east of the metropolitan area and attack into the built-up area after the crossing. Early tactical objectives include crossing points for additional combat and support units so as to minimize lines of communications lengths.

Planning is detailed and centralized. Full use is made of aerial photography, digital mapping, the **Big Cities** system, and satellite imagery. Rehearsals at the lowest levels are conducted in city streets and buildings under allied control; higher-level exercises utilize simulations and iterative computer runs to support course of action development and subsequent rehearsals. Unlike the Persian Gulf War of 15 years before, manpower shortages and human exhaustion, not fuel consumption, are identified as the likely causes of culmination during coming offensive operations. Historical examples reflect that friendly casualties will be high. Control of civilians and enemy prisoners of war will consume well beyond the strength of military police assigned to the theater; fortunately the Korean police have been undertaking training to assume these roles. However, soldiers will have to clear buildings of enemy. A lack of area and point denial munitions means they will also have to guard structures after they are cleared. Commanders doubt that sufficient manpower exists to accomplish all necessary tasks.

Priority intelligence requirements (PIR) at various levels identify the need for locating enemy headquarters and communications nodes, concentrations of civilians, the adversary's division reserve, and key facilities such as power plants, water treatment and pump stations, and charcoal stockpiles (charcoal still being an important heating fuel for many South Koreans). Intelligence officers are also told to rapidly detect any effort to reinforce the defenders. Additionally, ROKA/US commanders want to isolate and destroy enemy forces rather than have to fight them repeatedly during successive withdrawals through delaying positions; as such, defending forces must be acquired, monitored, and engaged either in their identified locations or when attempting to move. Micro-UAVs such as the "**Black Widow**" play a significant role in servicing these PIRs. Some are cycled so as to maintain constant surveillance of critical facilities, nodes, rooftops, and routes; their real-time feedback provides fire support units manning systems such as **EFOG-M** with a robotic forward observer capable of acquiring, monitoring, and providing battle damage assessment (BDA). Other **Black Widow** versions are flown around and into buildings in efforts to collect information in support of PIRs and other mission requirements, to include determination of indirect fire registration points using the UAV's onboard GPS capability.

Preliminary operations also include initial efforts to mold noncombatant behavior. Separation of civilians from enemy forces and other targets begins with the transmission of messages and other psychological operations initiatives.

The attack is initiated with employment of "**smart smoke**" to conceal crossing operations. The smoke is designed to deny the enemy visibility through the use of the naked eye and the vision enhancement systems known to be in his possession. Application of **smoke** to critical nodes, likely observation points, and identified air defense positions, in addition to its use on the river itself, adds to the adversary's isolation and inability to engage aircraft during insertions of soldiers and **MDARS-E robots**. Aerial platforms employing **high-energy acoustic** weapons systems attack headquarters and other targets to further this sense of disorientation and otherwise disrupt critical functions. EFOG-M engage selected targets to preclude an effective response to the allied effort.

After a successful river crossing, units dispense obstacle-clearing **robots** before advancing into the dense networks of city streets. Vision enhancement systems provide the attackers the ability to see through the **smart smoke** tailored to cause minimal interference with friendly goggles. **Black Widow** cameras help snipers acquire the most threatening targets during the crucial first moments as units wade ashore.

Comprehensive, up-to-date situational awareness has been identified as critical to operations in urban areas. The attacking force must know where the defenders are located, what weapons they have, where obstacles have been placed, and whether noncombatants are present. Information on the physical characteristics of the area of operations is also important, e.g., material regarding access routes, lines of sight, structural data, and key nodes such as water sources, power distribution facilities, and telephone switching centers. Components of this information must be promulgated and displayed to soldiers and leaders at each echelon.

Two capabilities should be demonstrated: (1) how to pass information quickly, completely, and securely in noisy urban areas with high levels of interference, and (2) how to display information in a manner that soldiers and commanders can effectively use.

The toughest C4I problems in urban areas involve signal degradation caused by interrupted lines of sight, extensive noise and jamming, multipath (especially at high frequencies), and requirements for significant bandwidth (especially problematic at low frequencies). Relays can alleviate some of these problems and can be demonstrated using (1) small, expendable devices able to pass voice, data, and video on multiple frequencies, and (2) airborne relays in the form of aerostats, UAVs, or LEO satellites.

A less obvious but still important problem is visualization of information. A useful tool, such as DIA's Big Cities system, can show terrain, force overlays, water and power nodes, and many other key data items in two and three dimensions. The system is accessible from the internet and, according to its designers, could be upgraded to project the impact of taking out key nodes or implementing special tactics. This tool could be physically demonstrated and implications of different levels of situational awareness shown. Alternatively, the impact of situation awareness can be examined during simulations by interactively presenting a human commander with changing levels of accuracy, timeliness, and completeness of information.

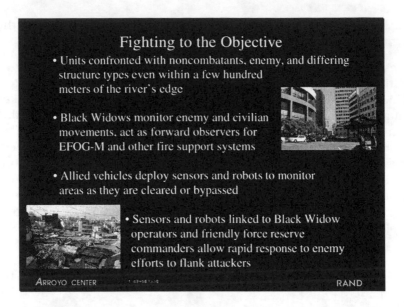

Units entering their first streets after gaining the north bank of the Han find themselves receiving fire, entering building types that range from modern architecture to densely packed homes with cinderblock walls, and they are beset with refugees greeting them as liberators. **Black Widows** and **MDARS-E** systems provide feedback of value to aerial engagement systems and snipers, but the difficulty of simultaneously monitoring present locations, dealing with ongoing events, and also filtering incoming intelligence from locally placed sensors mitigates the value of real-time inputs other than what is provided by a soldier's own senses and the shouts and signals of his fellow unit members. While less well-equipped ROK forces struggle with intermittent communications, improved radio systems provide U.S. soldiers continuous contact within units and with higher headquarters. These higher headquarters also find that they can accurately monitor friendly force locations, and in several instances staff personnel break into small unit nets to preclude potential fratricide incidents as soldiers move through the tangle of city streets, buildings, and alleyways.

Despite the dangers of overwhelming the small-unit leader with information, the advantages of UAV and UGV intelligence providers should not be underrated. Their accurate and timely input allows senior commanders to effectively isolate the attackers by denying the enemy movement of wheeled and mechanized/armored reserve forces. Foot-mobile enemy are more difficult to interdict, but prior knowledge of their initial locations and preemptive **EFOG-M** fires attrits even these threats.

The Objective

• 1 building, 16 floors, 120 apartments, 538 rooms

• Enemy air defense and height of building makes ground level entry the only viable option

• UAV/UGV systems support isolation of structure during clearing

• Pysops and high energy acoustic systems assist in controlling noncombatants and denying enemy use of civilians as shields/screens

• Thru-wall sensor system-of-systems speeds clearing operations and reduces loss of noncombatant life

ARROYO CENTER   4 472458 1 1 10   RAND

A typical battalion objective is shown here: a single high-rise building with several hundred rooms, each of which must be cleared. The structure's occupants include an unknown number of enemy and hundreds of noncombatants. The height of the building, the number of such targets, and enemy air defenses combine to make ground-level entry the only viable alternative (in exceptional circumstances, subterranean access may be available).

As is the case at the operational level, this target must be isolated so as to preclude enemy movement of personnel, munitions, or supplies in support of the defenders. Similarly, defending forces attempting to move to other positions outside the building must be interdicted so as to deny them the ability to later engage friendly forces. **Black Widow** and **MDARS-E** monitor the building exterior and assist in the employment of **EFOG-M** and other fires to accomplish these tasks. **Black Widow** also assists in detecting enemy positions via thru-window imagery and low-speed entry into the building where access is available. **Thru-wall radar** is used with other systems to detect human presence in rooms, corridors, and other spaces. While each system alone may at times be sufficient to determine whether one or more individuals occupy a given space, together they provide a means of determining with fair accuracy whether the occupants are enemy, civilian, or a combination of the two. In cases where civilians and combatants are intermixed, employment of **high energy acoustic weapons** incapacitates those in the target room or acts to separate combatants from innocents.

Intraunit **communications** allow small-unit leaders to minimize fratricide opportunities; contact with higher headquarters facilitates resupply, casualty evacuation, and status reporting. Headquarters continue to monitor subordinate unit locations and further restrict the chances of fratricide by ensuring that organizations are aware of other friendly forces in close proximity.

The reliability of **thru-wall systems**, use of foam denial, and the monitoring of access points via **Black Widow** and **UGVs** reduces the number of soldiers used as stay-behind security forces, but the maintenance of reaction forces tasks limited manpower resources. Soldier exhaustion requires the company commander to frequently rotate his platoons; **thru-wall systems** reduce the number of rooms that must be entered. This both speeds the clearing operation and reduces the risk of booby-trap and direct fire losses. Use of flash-bang and other nonlethal munitions also restricts the amount of debris soldiers confront, further limiting injuries during the operation. Direct-fire confrontations still take their toll during clearing and securing operations; however, missions that previously required battalions can now be undertaken by fewer soldiers. Though casualties are high, both friendly and noncombatant losses are fewer than would have been the case had rooms been entered blindly, had access to cleared areas remained open to the enemy, or had communications been interdicted by walls and other structures.

MOUT Strategic Concept:
Initial Observations

• Revised urban operations doctrine, derivative training, and advanced future technologies offer soldiers a solution to the currently unavoidable high costs of MOUT.

• Existent urban simulations require modifications for effective systems evaluation and analysis

ARROYO CENTER   RAND

Changes to doctrine and training and the seizing of opportunities offered by technological advances provide potential for upgrading the currently unacceptable levels of MOUT force readiness. In the longer term, however, there is a need in any but essential circumstances to remove the soldier from the very short ranges and density of fires inherent in urban combat. Initial analysis reflects that a reorientation from reliance on close combat operations to a doctrine of primarily offset operations will be feasible.

While historical evidence, common sense, and current technological capabilities all suggest that offset MOUT is viable, analytic tools that provide a forum for testing relevant concepts are greatly needed. No single existent simulation provides the scope and resolution necessary to accomplish these tasks, though a system of extant systems offers considerable potential in this regard.

Despite promising initiatives being pursued by several Army and USMC agencies, improvements in MOUT may be hindered by the present lack of a coordinating agency. The absence of an influential champion for all aspects of urban operations has delayed MOUT improvements within the Army. Worldwide demographics and U.S. strategic interests ensure continued commitment of American forces to urban environments; Bosnian operations, ongoing efforts in Haiti, and the 1992 Los Angeles riots are examples of such missions. Current capabilities and a demonstrated intolerance of friendly force casualties cast doubt on armed forces' abilities to support national objectives during urban operations contingencies.

Future MOUT Initiatives: One Path Forward

- Build on ACTD efforts to provide strategic continuity during the ACTD->Force XXI->AAN transition

- Modify simulations software to facilitate selection of best doctrine, training concepts, and technologies

- Produce quantitative system-of-system analyses for expanded ACTD, Force XXI, and AAN in all three areas

- Move from a doctrine reliant on close combat MOUT to an offset approach to urban operations

ARROYO CENTER   RAND

Pursuit of alternatives to close combat MOUT demands a thorough qualitative and quantitative analysis. The testing of specific technologies requires extensive modification of available simulations and considerable postquantitative analysis work. The linkage of strategic requirements to tactical and operational-level capabilities has been neglected. Consideration of a comprehensive continuum tying tactical MOUT requirements to strategic demands is essential for the ACTD, Force XXI, and AAN periods and beyond.

PRESENTATION BY LTC TONY CUCOLO

MOUT IN BOSNIA

EXPERIENCES OF A HEAVY TASK FORCE

DECEMBER '95 TO DECEMBER '96

LTC Tony Cucolo

My Bottom Line:

A post-settlement peace operation is
                    Precision MOUT"+".

Participation in these types of operations will increase
      in frequency.

We can do more to prepare our soldiers for this
      unique operational environment.

Agenda

Unique Aspects of Post-Settlement
    Peace Operations
Background Information
Life in the City and Discovery Learning
Wish List
Closing Thoughts

Aspects of Post-Settlement Environment

-- Post **intrastate** conflict environment:

* tensions high; great distrust
* probably no clear winner
* requirements to separate, disarm, de-mobe
  and canton forces
* scarcity of basic needs
* no law or there is rule by law
* someone is benefiting from the status quo

Aspects of Post-Settlement Peace Operations

★Soldiers, often first on the scene, find themselves
  active in third party roles...as mission essential
  tasks. Civilian and military requirements overlap
  and complement each other.

★It's construction, not destruction: soldiers become
  the enablers for reconstruction, resettlement,
  and establishing conditions for return to
  civil society. This will require applying varying
  degrees of force against belligerents who stand in
  the way of the settlement .

Brcko Primer

-- "Cleansed" by Serb para-military forces, Spring '92;
35,000 Bosniaks and Croats are expelled;
Serb population jumps from 8,000 to 31,000 by '95.

-- Undecided at Dayton: Left for arbitration.

-- For Bosnian Serbs: it is the East-West corridor linking
their Republic.

-- For Bosniaks and Bosnian Croats: the link to Croatia;
a commerce center for NE Bosnia; a key port on
on the Sava; former home to 35, 000.

Brcko Primer

City was home to:
-- RS Infantry Brigade
-- RS Special Police Detachment
-- RS Regional "Public Safety" HQ
-- 11 alleged war crimes sites

In Sector (upon arrival):
-- 20 RS Battalions
-- 6 Bosniak Battalions
-- 6 Bosnian Croat Battalions

Task Force 3-5
The Black Knights

-- Organized around the 3rd Battalion, 5th Cavalry Regt,
   a BFV battalion of 1st BDE, 1st Armored Division.

-- DEC 95 to AUG 96:
   3 Mech, 1 Tank, Arty Bttry, Radar, EN Co.

-- AUG 96 to DEC 96:
   4 Mech, 1 MP(ABN), Arty Bttry, Radar, EN Co.

-- Average Task Force strength was 1,100 soldiers.

Visual Examples
of the
Operating Environment

Precision MOUT "+" in Brcko: One Unit's Life in the City

-- Separate Forces: when trenches and bunkers are
         gone, and fortifications are not allowed,
         buildings become positions for faction
military.
         (secure an area/a border/an urban area,
         establish/maintain a zone of separation)

-- Ensure Freedom of movement: first, and always,
         for the force; then increasing demands for
         NGO and civilian freedom of movement
         (mine clearance, clear/secure a route,
         escort a convoy, react to sniper/ambush/
         indirect fire, react to civil disturbance)

Life in the City

-- Show presence: Establish lodgments, observation
         posts and checkpoints. Many of the best positions
         were in destroyed suburbs, around buildings.
         For eight months the unit secured an international
         crossing -- the Brcko Bridge. From these positions
         soldiers patrolled day and night at unpredictable
         times, and conducted random searches of vehicles,
         personnel and buildings.
         (mine clearance, conduct patrols, establish
         fighting positions, prepare a bunker,
         defend MOUT/building, secure and defend
         unit position, react to sniper, react to ambush)

Life in the City

-- Verify aspects of compliance.
   (conduct liaison/negotiate, search a building,
   process documents and equipment, demonstrate
   show of force, secure an urban area)

-- Enhance and Support the Rule of Law: CIVPOL slow
   to change; great potential for govt-sponsored
   crime, individual crime, and harassment and
   terrorizing of returnees; *PIFWCs*.
   (conduct patrols, demonstrate show of force,
   handle captured belligerents, conduct liaison/
   negotiate, apprehend/detain noncombatants,
   plan for media, react to press, cordon/search)

Discovery Learning

Support Elections. The scope and nature of election
   support was unexpected, but within the unit's
   capability. Tasks included: provide safe meeting
   sites, military security for registration and
elections;
   •   erect polling sites; secure and transport ballots;
   secure counting houses; transport ballots out of
   sector...
   (provide command and control, conduct patrols,
   secure an area/ an urban area, plan for media,
   react to press interview, secure a route, escort a
   convoy, defend MOUT/building, execute an
   R&S plan, employ a quick reaction force, and
   react to civil disturbance).

Discovery Learning

Provide security for persons and property with protected status.

-- UN personnel, low-profile NATO or State Dept personnel, NGOs, IPTF, ICTY, international observers and monitors and their dwellings, offices, vehicles and gear.

-- Requires careful planning and frequent rehearsal of: employ a QRF, perform infiltration/link-up/ exfiltration, execute an assault (MOUT), secure a building, and react to civil disturbance, and the "snatch" drill..

Discovery Learning

Defense and counter urban terrorism in a resettlement area.

-- Big sector, little unit.
-- Urban R&S plans. Building analysis.
-- Heavy use of: dogs, all available sensors, any and all available thermal/night vision equipment to include tank, BFV, and AVENGER FLIR, and NATO special ops units.
-- Snap checkpoints; aggressive searches.
-- Enemy adapted well.

Discovery Learning

<u>Crowd</u> <u>Control</u>.

IT IS AN UGLY THING TO BE ARMED
AND OUTNUMBERED.

-- Our only non-lethals were JET NOISE, ROTOR WASH,
BOREDOM and the BUTT STROKE.

-- Heavy use of liaison and contact with local leaders,
info operations, pre-emptive shows of force,
and the "block and channel" -- obstacles to
impede movement and snap checkpoints to
do individual search.

-- Combat multipliers: attitude, posture, professionalism
impartiality, zero tolerance and "The Switch."

Wish List

Soldier Protection: better body armor; pads
for the common man; position reinforcement.

A non-lethal that can stop a vehicle.

Street/road wide quick obstacles (detention).

Demo sniffing devices I don't have to feed.

More sensors; magnetic and seismic.

Individual restraints (or more zip cuffs!).

A politically acceptable riot control agent.

A small unit radio for the common man.

Closing Thoughts

-- Peace ops require readiness in precision MOUT.
   There is a general attitude of avoidance
   in dealing with fighting in built-up areas
   in heavy units -- and heavy units are very
   capable for these operations.

-- Doctrine and/or TTP needs consolidation.
   We went to a wide variety of manuals, TC's, pubs,
   and white papers to get TTP for training -- and
   in some cases execution -- of our required tasks.

Closing Thoughts

-- Training for high intensity conflict, then receipt of
   mission specific training prior to deployment
   works.

-- Well trained, disciplined troops cannot let mobs rule
   or urban terrorists jeopardize implementation
   of an agreement. We can better equip and train
   troopers heading into these unique environments
   if we accept the fact we will have to face these
   challenges.

PRESENTATION BY LTC GARY SCHENKEL

LtCol Gary
Schenkel
USMC

Sea Dragon 5 year Experimentation Plan

1996	1997	1998	1999	2000
		"URBAN WARRIOR" Operations in urban, near urban, and close terrain. Units conduct intelligence gathering, targeting, maneuver, and close combat; command/control by MEF (Fwd) CE		**"CAPABLE WARRIOR"** Operations combining virtual and live forces comprising operational level deception and maneuver in response to crisis, with objective of containing/obviating incipient MRC
"HUNTER WARRIOR" Small unit operations on a dispersed, open battlefield. Units conduct intelligence gathering and targeting, command/control by SPMAGTF (X) CE				

– Modular Command Element	– Non-lethal, to include C^2	TBD
– Engagement Coordination	– C4I	
– Non-lethals	– Sensors, to include C^2	
– Survivability	– Aviation Urban Ops	
– Sustainability	– Indirect Fires in Urban Ops	
– C4I	– Precision Targeting	
	– Position Location	
	– Mobility (to include aerial and subterranean)	
	– Unmanned Vehicles	

URBAN WARRIOR

Objective: Identify key capabilities necessary to conduct seabased, Naval expeditionary operations, in the unique environment of the urban littoral, from operations other than war, to mid-intensity conflict. Objectives include the ability to project Naval power into the littoral from a sea base, and to achieve battlespace dominance, across the spectrum, at the time, place and duration of our choosing. These operations must be conducted within the context of anticipated political and cultural restrictions.

URBAN WARRIOR

- **Cradle and test bed for development of:**
  - Operational concepts in the Urban Environment
- **T3P**
  - Tactics
  - Techniques
  - Technologies
  - Procedures

 URBAN WARRIOR

- **SEA DRAGON**
  - **Phase 1- Hunter Warrior**
    - Lateral, Dispersed, Noncontigious Battlefield
  - **Phase 2- Urban Warrior**
    - Lateral, Dispersed, Noncontigious Battlefield
    - + Vertical Dimension
    - + Subterranean
  - **Phase 3- Capable Warrior**

URBAN WARRIOR

- **THE 21ST CENTURY LITTORAL BATTLEFIELD**
  - 60–85% of world population in urban environment
  - 300 cities excess 1 million population
  - 80% of national capitals
  - 70% of nuclear reactors

 # URBAN WARRIOR

- **POPULATION MIGRATION**
  - Agrarian
  - Industrial
  - Information
- **URBANIZATION OF THE WORLD**
  - 70% of world population on littorals of the world
  - 85% by year 2015

URBAN WARRIOR

- **SOCIAL RESTRUCTURE**
  - Disregard for legal boundaries
  - Ethnic relocation / dislocation
  - Values
  - Expectations
- **SOCIAL UNREST**
  - Berlin / Heidelberg
  - London

URBAN WARRIOR

- **UNITED NATIONS**
  - **URBAN "AGGLOMERATIONS"**
    - **CONTIGUOUS DENSLEY POPULATED AREAS WITHOUT REGARD TO ADMINISTRATIVE BOUNDARIES**

URBAN WARRIOR

- **POPULATION EXAMPLES**

POPULATION	1900	1990
NEW YORK CITY	3,437,000	7,322,000
LOS ANGELES	102,000	3,485,000

URBAN WARRIOR

- **LAST 20 YEARS**
  - **LOS ANGELES, CALIFORNIA**
    - 3,485,000   39%
  - **SAO PAULO, BRAZIL**
    - 26,000,000   66%
  - **SEOUL, SOUTH KOREA**
    - 11,000,000   71%
  - **BOMBAY, INDIA**
    - 14,500,000   120%

URBAN WARRIOR

- **DECAY OF URBAN INFRASTRUCTURE**
  - ROAD SYSTEMS
  - SUBTERRANEAN
  - COMMUNICATIONS
  - UTILITIES
- **MORAL AND CULTURAL**
  - COMPLETE CHANGE IN TRADITIONAL VALUES

 URBAN WARRIOR

- **PHYSICAL STRUCTURE**
  HISTORICAL ECONOMIC SUCCESS
    GRANITE AND CONCRETE
  ECONOMICALLY STABLE
    BRICK AND WOOD
  URBAN SPRAWL
    TARPAPER AND CARDBOARD
- **INTENSE POPULATION DENSITY**

URBAN WARRIOR

- **HIGH INTENSITY**
  - **STALINGRAD, MANILA, HUE**
- **LOW TO MID INTENSITY**
  - **CHECHNYA**
  - **SARAJEVO**
- **CNN FACTOR**

URBAN WARRIOR

- **URBAN WEAPONS MIX**
  - **WEAPONS VS. STRUCTURES AND POPULATION**
  - **LETHAL VS. NONLETHAL**
  - **MOBILITY**
- **DIRECT FIRE VS. INDIRECT FIRE**
  - **AVIATION**
  - **AUTONOMOUS**

URBAN WARRIOR

- **URBAN PENETRATION**

- **URBAN THRUST**

- **URBAN SWARM**

 # URBAN WARRIOR

- **HUNTER WARRIOR**
  - **ENGAGEMENT AT EXTREME DISTANCE**

- **URBAN WARRIOR**
  - **AVG POINT OF CONTACT 35M**
  - **FREQUENT POC   6IN TO 15M**
  - **EXTREME POC   150M**

URBAN WARRIOR

- **THE EXPECTED ENEMY**
  - **ASYMETRICAL**
  - **COMBINATION STATE SUPPORTED, UNIFORMED, PARAMILITARY**
- **EXTREME**
  - **TERRORIST TYPE**
  - **NONUNIFORMED**
  - **TECHNOLOGICALLY EQUAL / SUPERIOR**
  - **WMD (CHEM BIO)**

URBAN WARRIOR

- **THE THREE BLOCK WAR**

  - **HUMANITARIAN ASSISTANCE**

  - **LOW INTENSITY CONFLICT**

  - **ALL-OUT COMBAT**

URBAN WARRIOR

- **LEVERAGING THE EXPERTS**
  - TREXPO
  - URBAN WARFARE CONFERENCES
  - CROSSCANYON MOBILITY
  - MARFORRES
  - SOTG
  - FBI
  - LOCAL LAW ENFORCEMENT

URBAN WARRIOR

- **SCOPE AND VISION**
  - MARINE AIR GROUND TASK FORCE
  - 170 THOUSAND US MARINES
  - OPERATE AROUND THE MARINE RIFLE SQUAD
  - RARELY IF EVER IN ISOLATION

URBAN WARRIOR

- **EXPERIMENT SCHEDULE**
  - **LOE 1, JANUARY 1998**
  - **LOE 2, APRIL 1998**
  - **LOE 3, JULY 1998**
  - **CPE, SEPTEMBER 1998**
- **URBAN WARRIOR AWE**
  - **SAN FRANCISCO BAY AREA**

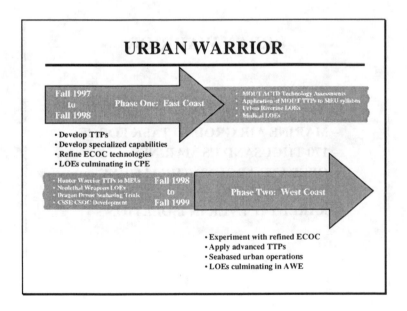

URBAN WARRIOR

[He] "felt a growing confidence" as he moved forces into the engagement area near the downed U.S. helicopter and established ambushes along likely United Nations relief force avenues of approach: "This claustrophobic battleground, in Aideed's stronghold, was where Aden had hoped to fight. Other militia platoons, he knew, would be rushing from the north, south, and east. The Americans were not supermen. In these dusty streets, where combat was reduced to rifle against rifle, they could die as easily as any Somali"

USN Proceedings

 # URBAN WARRIOR

- **CITY EXCESS 1 MILLION POPULATION**
- **3RD WORLD TYPE**
  - DETACHED, SEMIDETACHED, 1 TO 3 STORY
  - SMALL CITY CENTER
    - GRANITE AND CONCRETE
    - NOT TO EXCEED 15 STORIES

URBAN WARRIOR

- **CORE ISSUES**
  - **REDUCTION OF MANPOWER REQUIREMENTS TO CONDUCT URBAN OPERATIONS**
  - **REDUCTION OF COMBATANT AND NON-COMBATANT CASUALTIES**
  - **AVIATION INTEGRATION**
  - **USE OF DIRECT / INDIRECT FIRES**
  - **LETHAL AND NONLETHAL**

URBAN WARRIOR

- **CORE ISSUES CONTINUED....**
  - **POSITION LOCATION / COMMUNICATION / C4I IN THE URBAN ENVIRONMENT**
  - **UNMANNED VEHICLE SYSTEMS**
    - **SURFACE**
    - **SUBSURFACE**
    - **SUBTERRANEAN**
  - **SENSORS**
  - **ADVANCED TACTICAL AND OPERATIONAL CONCEPTS COUPLED WITH ENABLING TECHNOLOGIES**

URBAN WARRIOR

- **MISSION PROFILES**
  - **EVACUATE U.S. CITIZENS**
  - **SEPARATE ARMED FORCES**
  - **DESTROY AN ORGANIZED / DISORGANIZED FORCE**
  - **POLICE GENERAL DISORDER**
  - **DISARM FORCES**
  - **POLICE DAILY**
  - **REBUILD AND OR RESTORE ORDER**

URBAN WARRIOR

OPERATIONAL FRAMEWORK

- **PENETRATE**   **P**
- **ISOLATE**   **I**
- **COLLAPSE**   **C**
- **SUSTAIN**   **S**
- **INFLUENCE**   **I**
- **CONTROL**   **C**

URBAN WARRIOR

"THIS IS AN EXPERIMENT"

GENERAL C.C. KRULAK

URBAN WARRIOR

- **SUSTAINMENT CHALLENGE**
- **COMBATANT AND NONCOMBATANT**
  - **HOST NATION**
  - **EXPEDITIONARY FINANCE**
  - **EXPEDITIONARY MEDICINE**

FROM THE SEA

PRESENTATION BY DR. RANDALL STEEB

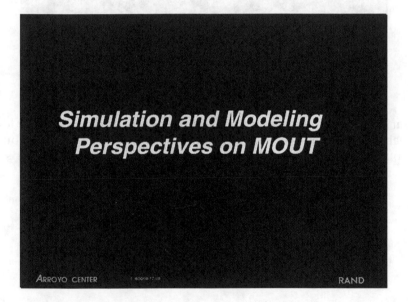

This briefing summarizes our recent review of the maturity and applicability of constructive simulations for MOUT. All of the simulations examined are either now available or are planned for release in the near future.

Our main focus was on examining simulations that can be used for analytic purposes—to explore and quantitatively evaluate new concepts and technologies for MOUT.

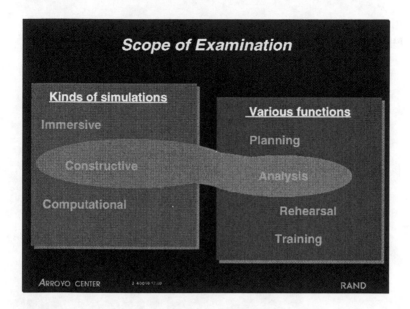

There are many types of MOUT-oriented simulations and models. These include three primary forms, which we term immersive, constructive, and computational. Immersive simulations attempt to replicate much of the experience of warfighting, and typically have real-time perspective scene generation, battlefield sounds, and realistically configured crew stations. Examples are distributed interactive simulation (DIS), advanced distributed simulation (ADS), aircraft simulators and trainers, and other virtual reality systems. Constructive simulations are more focused on wargaming representation, and they typically have map-oriented graphics and non-real-time operation. These are usually designed for efficient execution of large numbers of excursions. Computational models, finally, tend to be specialty programs that calculate outcomes—weapons effects, sensor performance, communications quality, etc.—but do not support time- or event-stepped simulations. Our examination focuses on constructive simulations, because these appear to provide the most potential for answering key questions about MOUT.

We also concentrate our examination on analysis, as opposed to planning, training, and rehearsal. Analysis is essential for determining the value of the candidate MOUT technologies and tactics. The other functions are also important, but are secondary to our examination. Some of these functionalities may be accomplished by the analytic models as an added benefit.

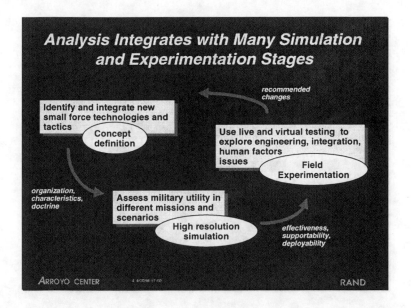

High-resolution simulation, as embodied in constructive models, is part of a multistage process for identifying and developing new concepts and technologies. The concept development stage acts as an input to the simulation process, specifying systems, performance parameters, and organizations to be examined. These are assessed in the simulation, and those with promise are passed on to the field experimentation phase, in which complex and time-consuming live and virtual testing takes place. Constructive simulation acts to filter down the large number of possibilities to a manageable number for field experimentation. All of these stages interact repeatedly during development of new systems and concepts.

For a simulation to be appropriate as an analysis tool, many aspects must be included. The scale must be sufficient to capture force sizes, terrain extent, and town sizes of interest. At the same time, there must be enough modeling resolution to differentiate key aspects (e.g., a soldier exposing just his head or his entire upper body to enemy fire). Scenario development must be straightforward and allow scripting of complex behaviors, such as running between buildings, calling for fire, and blowing entry holes in walls. The technical characteristics of sensors, radios, platforms, weapons and other devices must be faithfully represented, including such aspects as weapon effects and communications multipath. Human factor issues have been largely ignored in simulations, but they can be extremely important in urban areas. The models should reflect impacts of fatigue, delayed reaction times, and other physiological factors. User interfaces and postprocessing, finally, should allow the user to be able to quickly set up excursion matrices and call for special measures of effectiveness.

> ## *What Kinds of Questions Can We Expect to Answer?*
>
> - Performance contributions: *Increase in situational awareness with micro-sensors and platforms?*
> - Requirements: *How fast must a UGV sprint to be survivable? How much delay will render communications unusable?*
> - System trades: *Armor vs. active protection, weapon footprint vs. target location error, etc.*
> - Mix of technologies and tactics: *What is the best combination of sensors, obscurants, and weapons and how should they be fought?*
> - Robustness: *What enemy responses are possible?*
> - Operational applicability: *What range of missions are feasible for an advanced dismounted force? for a future combined arms force?*
>
> ARROYO CENTER   RAND

This is a list of increasingly difficult questions that the simulations may be called on to answer. The first and most straightforward question is how much a candidate system or tactic contributes to force performance. An example is the added coverage area and increased number of contacts provided by a force equipped with seeded microsensors, compared to a baseline force. Somewhat more difficult is the determination of minimum levels of performance to achieve a mission. This requires determination of success or failure across a range of conditions and threats. Additional modeling is required for system trades, whether between similar system types or among completely different ways to achieve a goal. Here the measures of effectiveness may range over such dimensions as deployability, lethality, survivability, supportability, and flexibility.

The last three questions are typically too hard to answer in an absolute sense. The simulations cannot be expected to determine the optimum mix of equipment and tactics for a force, its robustness against all forms of threats, or the complete range of missions in which the force will be effective. But the simulation should provide insights that can guide further research, development, and testing.

Four Simulations Were Examined

- **JANUS and its associated suite of models**
  - Developed at LLNL
  - Transitioned to U.S. Army (TRAC)
  - Extended and augmented at RAND
- **CAEN (Close Action Environment Model)**
  - Developed at CDA (UK)
  - Ported by Australian MOD
- **JCATS (Joint Conflict and Tactical Simulation)**
  - Developed at LLNL by merging JTS and JCM
  - Precursors in use by Army, Marines, AF, DOE
- **IUSS (Integrated Unit Simulation System)**
  - Developed by US Army Natick Lab
  - Special purpose program--operates on PC

ARROYO CENTER   RAND

We examined four promising constructive simulations for MOUT analysis. The first, JANUS, has been in operation at RAND for almost ten years. During this period it has been strongly modified and augmented for analysis rather than training applications. CAEN was also originally based on JANUS, and is a MOUT-specific simulation developed by the Centre for Defense Analysis in the U.K. The current simulation is written in Turbo-Pascal but has been ported to Symbolics machines, and it will soon be ported to Suns. JCATS is a soon-to-be-available combination model with the best features of JTS and JCM, which have been in extensive use with many agencies. IUSS, finally, is a special-purpose physiological modeling tool, which can supplement the other models.

Two Example Scenarios Illustrate Potential Contributions

- Establish corridor and move convoy through enemy-held town
  - Primarily external to buildings
  - Localized combined arms engagement
  - Focus on mobility, sensor-shooter links
- Clear urban core area of entrenched forces
  - Adds interior fighting, movement from building to building, larger scale engagement
  - Increased problems with rubbling , fratricide, non-combatants, coordination

ARROYO CENTER   RAND

We considered two different example scenarios to illustrate the characteristics of the several simulations. The first of the two was smaller in scope and primarily exterior to the buildings. Here the Blue force is in convoy formation, has limited weapons, and attempts to move through streets flanked by enemy-held buildings. The second scenario is the large-scale effort in Seoul described in the conference handouts. This includes storming buildings, fighting through rooms, and being concerned about noncombatants.

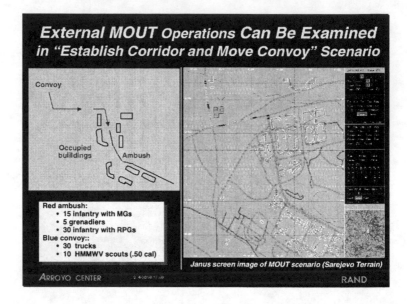

Janus screen image of MOUT scenario (Sarejevo Terrain)

The example convoy vignette is shown in the JANUS screen image above. Blue is escorting a resupply or humanitarian convoy of trucks through the downtown area. It leads with HMMWV-scouts equipped with .50 caliber machine guns, and it changes routes if an enemy ambush is spotted in time. Red has prepared an ambush partway through the town, with cratering charges along the road and infantry in the nearby buildings. Red waits until most of the convoy is in the killing zone and opens fire. Preliminary runs show that the Red is typically able to achieve surprise, the lead vehicles are hit, the convoy is halted, and most of the remaining vehicles are destroyed.

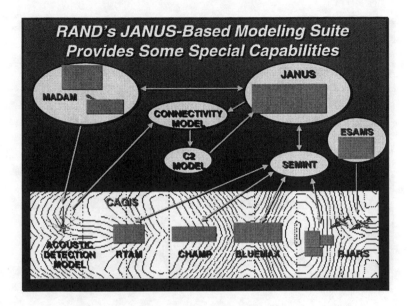

RAND's JANUS-based ensemble of locally distributed models has evolved over the last 10 years to include specialty representations for smart munitions, command and control, air defenses, acoustic sensors, and low observables. Some of these models are also applicable to MOUT operations, but most were developed for modeling future operations in open and close terrain.

What JANUS Suite of Models Provides

- Basic situation: building shells, streets, firing ports, lines of sight
- Mobility representation: simple vehicle and dismount movements, more complex air maneuver
- Engagement dynamics (sensing, communication linkage, coordination, firing, losses, simple suppression)
- Partial sensor blinding from smoke, weather
- Limited responsive threat reactions to Blue operation
- interactive and batch modes of operation

ARROYO CENTER   11 4/30/96 17.00   RAND

JANUS-based simulation, as it is employed at RAND, models only a portion of MOUT engagements and in fact is not well suited to modeling interior fighting. It does model exterior fighting to a reasonable extent. Units may fire from buildings on soldiers and vehicles in streets, open areas, and behind cover. The attackers may return fire, move, dispense smoke, or take other actions. Many different excursions—use of UGVs, mounted and dismounted operations, and equipping the force with different weapons—were run in the "establish corridor" scenario. The larger-scale Seoul scenario would be much more difficult to represent successfully.

In general, the strengths of JANUS-based constructive modeling are in the system-level actions it can represent and assess. It can be used to determine the contributions of better sensors, faster movement, longer-range and higher-precision weapons, different types of smoke, and various threat reactions.

The JANUS-based suite of models provides a two-dimensional map view of the combat situation and, like many constructive models, does not represent many of the special aspects of MOUT. For example, the scene illustrated in this slide is a rendering of Copehill Down in the U.K., one of the premier MOUT training sites in the world. The 3-D view shows the importance of building structure, look-down aspects from windows and firing ports, sloped roofs, and many other issues that JANUS does not address.

Even so, a 3-D rendering may not be sufficient to model all of the critical aspects of urban areas. For example, the above photograph of one of the Copehill Down roads shows the complexity of fighting. Training exercises often use underground areas covered by logs and rubble. These may be occupied shortly after troops have passed; then the troops are engaged from the rear, with the snipers moving to the next position after taking a few shots. Again, this is very difficult to model with simulation.

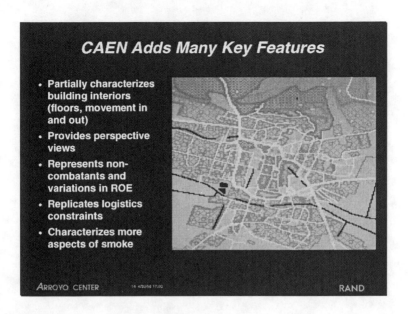

CAEN makes a nice start at modeling more aspects of urban operations because it represents portions of the building interiors, shows 2-D and 3-D views from different locations, and models more special aspects of MOUT. It is somewhat limited in scope, however, as it currently can display only a 5 × 5 km area, and forces are constrained to company size and below.

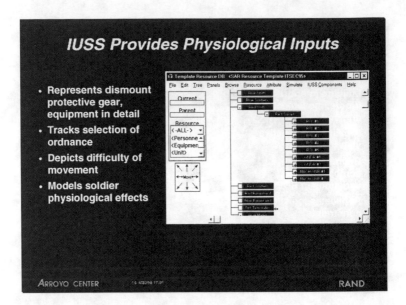

IUSS is a specialty, PC-based model that primarily gauges the effects of exertion and protective equipment on soldier physiological performance. The movements and conditions from a larger simulation such as JANUS or CAEN can be input to IUSS for an assessment of the movement speed and soldier capability. These outputs can be returned to the larger simulation or used independently in off-line analyses.

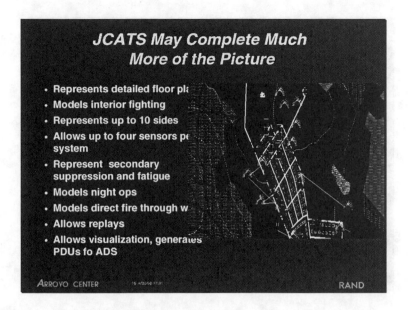

The Joint Combat and Tactical Simulation, now nearing its release date, may provide the greatest amount of scale and resolution of any of the simulations examined. Unlike CAEN and JANUS, it is able to model interior fighting, with representations of floors, walls, interior doors, and many other building characteristics. Enemies, friendlies, and noncombatants can have many different affiliations. While JANUS equips each entity with two sensors maximum, JCATS allows as many as four. Areas as large as 600 km on a side can be played, with as many as 20,000 entities present. Suppression is projected to include both the standard forms of JANUS and CAEN, but also a secondary suppression in which the entity moves to a hide posture. There is some discussion about how much 3-D visualization and replay will be present, but the user should be able (as in CAEN) to call up perspective images from different vantage points. The model is expected to be in beta test form at the end of March 1998, and ready for release a month or so later.

> ### *Desired Measures of Effectiveness*
>
> - **Detections: completeness, redundancy, coverage and accuracy**
> - **Timeliness: delays for detection, communications, decisionmaking, response**
> - **Area controlled: occupied, denied, safe; nodes neutralized**
> - **Attrition: kills and losses (including fratricide and non-combatants), infrastructure damage, shock effects, fatigue**
> - **Attribution: who was responsible for each outcome**
> - **Participation: what systems were active**
>
> ARROYO CENTER   17 4/35/96 17:H   RAND

Measures of effectiveness are similar to those in conventional warfare, with some aspects specific to urban fighting. Detections will have to have greater accuracy and completeness for effective fighting in urban areas compared to open areas. Time delays will have to be very short due to the short ranges and short exposure times in urban operations. Area controlled will be weighted by the importance of nodes and key areas in cities. Attrition must include noncombatants, friendlies, and collateral damage. Finally, the outcome criteria in a MOUT operation may be different than the "go to ground," penetration of lines, control of an area, or disintegration outcomes seen with conventional, open area warfare. It may be sufficient to isolate an area, release hostages, bypass a town, or control key nodes.

Many Aspects Still Need Resolution

- Noise, confusion, and disorientation
- Communications degradation--multipath, interference, contention, jamming
- Effect of clutter and obscurants on detection
- Richochet, rubbling, accidents
- Localized field of view--attention directed to near field
- Influences of shock, momentum, isolation
- Mission creep, change of sides by non-combatants
- Weapon accuracy while running, diving, climbing
- Interactions between new weapons, e.g. acoustic sidelobes on weapon aiming and smoke persistence

ARROYO CENTER   RAND

The constructive models examined provide a basis for answering many key questions about MOUT operations, but there are still many aspects that currently appear to be too hard to emulate. Among these are many of the issues brought up at MOUT training sites—noise and confusion from weapons and obscurants, rapid changes in mission, and problems with communications and position-location systems. Some of these aspects may be captured in DIS systems, but the processing power and interaction levels required may be prohibitive. The most cost-effective approach may be to determine the impacts of many of these effects off-line from DIS systems and field tests and feed the resulting factors (reduced reaction times, reduced accuracy, etc.) into constructive simulations.

BIBLIOGRAPHY

Chuikov, Vasili I., *The Battle for Stalingrad,* New York: Holt, Rinehart, and Winston, 1964.

von Clausewitz, Carl, *On War,* Michael Howard and Peter Paret (eds., trans.), Princeton, NJ: Princeton University Press, 1976.

Donnelley, Thomas, Margaret Roth, and Caleb Baker, *Operation Just Cause: The Storming of Panama,* New York: Lexington, 1991.

Hammel, Eric, *Fire in the Streets: The Battle for Hue, Tet 1968,* New York: Dell, 1991.

Larsen, Eric, *Casualties and Consensus: The Historical Role of Casualties in Domestic Support of U.S. Military Operations,* Santa Monica, CA: RAND, MR-726-RC, 1996.

Report of the Defense Science Board Task Force on Military Operations in Built-up Areas (MOBA), November 1994.

Thomas, Timothy L., "The Caucasus Conflict and Russian Security: The Russian Armed Forces Confront Chechnya III. The Battle for Grozny, 1–26 January 1995," *Journal of Slavic Military Studies,* Vol. 10 (March 1997), pp. 50–108.